The Green Energy Shift

Unlocking the Potential of Renewable Resources for a
Cleaner Planet

Luna Harmon

1

The Green Energy Shift

© Copyright 2024 by **Luna Harmon**

All rights reserved

TABLE OF CONTENTS

Chapter 1: Solar Energy: Leading the Charge

Fundamentals of Solar Power and Its Role in the Shift

Solar power stands as a beacon of hope in the quest for sustainable energy solutions. Its potential to transform the global energy landscape is immense, driven by the sun's inexhaustible supply of energy. Understanding the fundamentals of solar power is crucial to appreciating its role in the broader shift towards renewable energy sources.

At its core, solar power harnesses the energy emitted by the sun and converts it into usable electricity. This process primarily involves photovoltaic (PV) cells, which are the building blocks of solar panels. These cells are made from semiconductor materials, typically silicon, that absorb sunlight and release electrons. This flow of electrons generates an electric current, which can be harnessed for power. The efficiency of this conversion process has improved significantly over the years, thanks to advancements in technology and materials science.

The appeal of solar power lies in its abundance and accessibility. The sun delivers more energy to the Earth in an hour than the entire world consumes in a year. This staggering statistic underscores the potential of solar energy to meet global energy demands. Unlike fossil fuels, which are finite and concentrated in specific regions, solar energy is available everywhere the sun shines. This universality makes it an attractive option for both developed and developing countries, offering a path to energy independence and security.

The environmental benefits of solar power are equally compelling. Unlike coal, oil, or natural gas, solar energy production does not emit greenhouse gases or other pollutants. This makes it a clean and sustainable alternative that can significantly reduce the carbon footprint of energy production. As concerns about climate change and environmental degradation mount, the shift to solar power becomes not just desirable but imperative.

Economic factors also play a significant role in the adoption of solar power. The cost of solar panels has plummeted over the past decade, making solar energy more affordable than ever. This decline in cost is largely due to economies of scale, technological advancements, and increased competition in the solar industry. As a result, solar power is now competitive with, and in some cases cheaper than, traditional energy sources. This economic viability is driving widespread adoption, from residential rooftops to large-scale solar farms.

The integration of solar power into the energy grid presents both opportunities and challenges. On the one hand, solar energy can reduce reliance on fossil fuels and enhance energy security. On the other hand, the intermittent nature of solar power—dependent on weather and daylight—poses challenges for grid stability and reliability. To address this, advancements in energy storage technologies, such as batteries, are crucial. These technologies allow excess solar energy to be stored and used when the sun is not shining, ensuring a steady and reliable power supply.

Solar power also plays a pivotal role in decentralizing energy production. Traditional energy systems are often centralized, with power generated at large plants and distributed over vast networks. In contrast, solar power can be generated locally, at the point of use. This decentralization reduces transmission losses, enhances energy security, and empowers individuals and communities to take control of their energy needs. It also opens up opportunities for innovative business models, such as community solar projects and peer-to-peer energy trading.

The role of government policy and incentives cannot be overstated in the promotion of solar power. Many countries have implemented policies to encourage the adoption of solar energy, including tax credits, subsidies, and feed-in tariffs. These measures help offset the initial cost of solar installations and make solar power more accessible to a broader audience. Additionally, governments are investing in research and development to drive further innovations in solar technology and integration.

Education and awareness are also critical components of the solar power shift. As more people become aware of the benefits and potential of solar energy, demand continues to grow. Educational initiatives, from school programs to community workshops, play a vital role in spreading knowledge and encouraging adoption. By understanding how solar power works and its benefits, individuals are more likely to invest in solar solutions for their homes and businesses.

The future of solar power is bright, with ongoing research and development promising even greater efficiencies and applications. Innovations such as bifacial solar panels, which capture sunlight on both sides, and solar skins, which allow panels to blend seamlessly with rooftops, are just a few examples of the exciting advancements on the horizon. As technology continues to evolve, the potential for solar power to transform the energy landscape becomes increasingly tangible.

In summary, solar power is a cornerstone of the shift towards renewable energy. Its abundance, environmental benefits, and economic viability make it a key player in the transition to a sustainable energy future. By understanding the fundamentals of solar power, we can better appreciate its role in this transformative shift and work towards a cleaner, greener world.

Technological Innovations Driving Solar Adoption

Technological innovations have been pivotal in accelerating the adoption of solar energy, transforming it from a niche alternative to a mainstream power source. These advancements have not only improved the efficiency and affordability of solar technology but have also expanded its applications, making it accessible to a wider audience.

One of the most significant breakthroughs in solar technology is the development of high-efficiency photovoltaic (PV) cells. Traditional silicon-based solar cells have seen remarkable improvements in their ability to convert sunlight into electricity. Researchers have pushed the boundaries of efficiency, achieving rates that were once thought unattainable. This has been accomplished through innovations such as passivated emitter and rear cell (PERC) technology, which enhances the light absorption and reduces energy loss. As a result, modern solar panels can generate more power from the same amount of sunlight, making them more cost-effective and appealing to consumers.

Beyond silicon, the exploration of alternative materials has opened new avenues for solar technology. Perovskite solar cells, for instance, have garnered significant attention due to their potential for high efficiency and low production costs. These cells are made from a unique crystal structure that can be manufactured using simpler and cheaper processes compared to traditional silicon cells. Although still in the research phase, perovskite cells promise to revolutionize the solar industry by offering a more affordable and efficient option for harnessing solar energy.

Another area of innovation is the development of bifacial solar panels. Unlike conventional panels that capture sunlight on one side, bifacial panels are designed to absorb light from both sides. This allows them to take advantage of reflected sunlight from the ground or surrounding surfaces, increasing their overall energy output. Bifacial panels are particularly effective in environments with high albedo, such as snowy or sandy regions, where the reflection of sunlight is significant. This technology not only boosts efficiency but also enhances the versatility of solar installations.

The integration of solar technology into everyday materials has also seen exciting advancements. Building-integrated photovoltaics (BIPV) represent a seamless way to incorporate solar energy into the architecture of buildings. BIPV systems replace conventional building materials with solar panels, allowing structures to generate their own electricity without compromising aesthetics. This approach is particularly appealing for urban environments where space is limited, as it transforms rooftops, facades, and windows into energy-generating surfaces.

In addition to hardware innovations, advancements in software and data analytics have played a crucial role in optimizing solar energy systems. Smart inverters and energy management systems enable real-time monitoring and control of solar installations, ensuring they operate at peak efficiency. These technologies allow users to track energy production, consumption, and storage, providing valuable insights for optimizing energy use. Furthermore, predictive analytics can forecast energy generation based on weather patterns, helping to balance supply and demand more effectively.

Energy storage solutions have also seen significant progress, addressing one of the primary challenges of solar energy: its intermittent nature. The development of advanced battery technologies, such as lithium-ion and flow batteries, has made it possible to store excess solar energy for use during periods of low sunlight. This capability enhances the reliability and stability of solar power, making it a more viable option for both residential and commercial applications. As storage costs continue to decline, the integration of solar and storage systems is becoming increasingly common, further driving the adoption of solar energy.

The rise of decentralized energy systems has been facilitated by innovations in microgrid technology. Microgrids are localized energy networks that can operate independently or in conjunction with the main grid. They are particularly beneficial in remote or underserved areas where traditional grid infrastructure is lacking. By incorporating solar panels and energy storage, microgrids provide a reliable and sustainable energy solution, empowering communities to achieve energy independence.

The solar industry has also benefited from advancements in manufacturing processes, which have reduced production costs and increased scalability. Automation and precision engineering have streamlined the production of solar panels, making them more affordable and accessible. This has been complemented by improvements in supply chain logistics, ensuring that solar technology can reach even the most remote regions.

Public policy and incentives have played a supportive role in fostering technological innovation in the solar sector. Governments around the world have implemented measures to encourage research and development, such as grants, tax credits, and subsidies. These initiatives have spurred investment in solar technology, driving further advancements and accelerating adoption.

As solar technology continues to evolve, the potential for new applications and innovations remains vast. Emerging technologies, such as solar skins that mimic the appearance of traditional roofing materials, and transparent solar panels that can be integrated into windows, are just a glimpse of what the future holds. These innovations promise to make solar energy even more ubiquitous, seamlessly integrating it into our daily lives.

In conclusion, technological innovations have been instrumental in driving the adoption of solar energy, making it more efficient, affordable, and versatile. From high-efficiency PV cells to advanced energy storage solutions, these advancements have expanded the possibilities for solar power, paving the way for a sustainable energy future. As research and development continue to push the boundaries of what is possible, solar energy is poised to play an increasingly central role in the global energy landscape.

Overcoming Challenges in Solar Energy Implementation

Implementing solar energy on a broad scale presents a myriad of challenges, each requiring innovative solutions and strategic planning. While the benefits of solar power are undeniable, the path to widespread adoption is fraught with obstacles that must be navigated to ensure successful integration into the energy landscape.

One of the primary challenges in solar energy implementation is the initial cost of installation. Although the price of solar panels has decreased significantly over the years, the upfront investment remains a barrier for many individuals and businesses. This is particularly true for low-income households and small enterprises that may lack the capital to invest in solar technology. To address this, various financing options have emerged, such as solar leasing and power purchase agreements (PPAs), which allow consumers to adopt solar energy with little to no initial cost. These models enable users to pay for solar power as they consume it, often at a rate lower than traditional electricity bills, making solar energy more accessible.

Another significant hurdle is the intermittency of solar power. Solar energy production is dependent on sunlight, which means it can fluctuate based on weather conditions and time of day. This variability poses challenges for grid stability and reliability, as energy supply must consistently meet demand. To mitigate this issue, advancements in energy storage technologies are crucial. Batteries and other storage solutions can store excess solar energy generated during peak sunlight hours for use when production is low. This not only ensures a steady power supply but also enhances the resilience of solar energy systems.

The integration of solar power into existing energy grids presents its own set of challenges. Traditional grids were not designed to accommodate the decentralized nature of solar energy, which is often generated at the point of use. This requires significant upgrades to grid infrastructure to manage the flow of electricity from multiple sources. Smart grid technology offers a solution by enabling real-time monitoring and control of energy distribution. These systems can dynamically balance supply and demand, optimize energy use, and reduce transmission losses, facilitating the seamless integration of solar power into the grid.

Land use and site selection are also critical considerations in solar energy implementation. Large-scale solar farms require significant land area, which can lead to conflicts with agricultural use, conservation efforts, and local communities. Careful planning and stakeholder engagement are essential to identify suitable sites that minimize environmental impact and address community concerns. Innovative approaches, such as agrivoltaics, where solar panels are installed above crops, offer a way to combine solar energy production with agricultural activities, maximizing land use efficiency.

Regulatory and policy barriers can impede the adoption of solar energy. Inconsistent regulations, permitting processes, and grid connection standards can create uncertainty and delay project development. Streamlining these processes and establishing clear, supportive policies are vital to encourage investment and expedite solar energy projects. Governments play a crucial role in setting ambitious renewable energy targets, providing incentives, and creating a favorable regulatory environment that fosters solar energy growth.

Public perception and awareness also influence the success of solar energy implementation. Misconceptions about the reliability, cost, and environmental impact of solar power can hinder adoption. Education and outreach efforts are essential to inform the public about the benefits and feasibility of solar energy. By showcasing successful projects and providing transparent information, stakeholders can build trust and support for solar initiatives.

The availability of skilled labor and expertise is another factor that can impact solar energy projects. The design, installation, and maintenance of solar systems require specialized knowledge and skills. Investing in workforce development and training programs is crucial to ensure a pool of qualified professionals who can support the growing solar industry. This not only facilitates the implementation of solar projects but also creates job opportunities and stimulates economic growth.

Technological advancements continue to play a pivotal role in overcoming challenges in solar energy implementation. Research and development efforts are focused on improving the efficiency, durability, and cost-effectiveness of solar technologies. Innovations such as flexible solar panels, which can be integrated into a variety of surfaces, and solar tracking systems, which optimize panel orientation to maximize energy capture, are expanding the possibilities for solar energy use.

Collaboration and partnerships are key to addressing the multifaceted challenges of solar energy implementation. Stakeholders from government, industry, academia, and civil society must work together to share knowledge, resources, and best practices. Collaborative efforts can drive innovation, streamline processes, and create synergies that accelerate the transition to solar energy.

In conclusion, while the path to solar energy implementation is not without its challenges, these obstacles are surmountable with strategic planning, technological innovation, and collaborative efforts. By addressing financial, technical, regulatory, and social barriers, we can unlock the full potential of solar energy and pave the way for a sustainable energy future. The journey may be complex, but the rewards of a cleaner, more resilient energy system are well worth the effort.

Case Studies: Successful Solar Projects Worldwide

Solar energy has emerged as a transformative force in the global energy landscape, with numerous projects around the world showcasing its potential to provide clean, sustainable power. These case studies highlight the diverse applications and successes of solar energy, offering valuable insights into how different regions have harnessed the sun's power to meet their energy needs.

One of the most notable examples of successful solar implementation is found in Germany, a country that has long been a leader in renewable energy. The Energiewende, or "energy transition," is Germany's ambitious plan to shift from fossil fuels to renewable energy sources. Central to this initiative is the deployment of solar power. Germany's commitment to solar energy is evident in its vast solar farms and widespread rooftop installations. The country has invested heavily in solar technology, supported by favorable policies and incentives that have encouraged both residential and commercial adoption. As a result, Germany has become one of the world's largest producers of solar energy, significantly reducing its carbon emissions and setting a benchmark for other nations to follow.

In India, the Kamuthi Solar Power Project stands as a testament to the country's dedication to expanding its renewable energy capacity. Located in the southern state of Tamil Nadu, this massive solar farm covers an area of over 2,500 acres and has a capacity of 648 megawatts, making it one of the largest solar power plants in the world. The project was completed in a record time of just eight months, demonstrating India's commitment to rapidly scaling up its solar infrastructure. The Kamuthi project not only provides clean energy to millions of people but also contributes to India's goal of achieving 100 gigawatts of solar capacity by 2022. This ambitious target is part of India's broader strategy to combat climate change and reduce its reliance on coal.

In the United States, the Solar Star project in California exemplifies the potential of large-scale solar installations. With a capacity of 579 megawatts, Solar Star is one of the largest solar farms in the country. It spans over 3,200 acres and consists of more than 1.7 million solar panels. The project generates enough electricity to power approximately 255,000 homes, significantly contributing to California's renewable energy goals. Solar Star's success is attributed to the state's supportive policies, abundant sunshine, and innovative technology, which have made California a leader in solar energy production.

Australia's solar energy journey is highlighted by the success of the Nyngan Solar Plant in New South Wales. This 102-megawatt solar farm is one of the largest in the Southern Hemisphere and plays a crucial role in Australia's transition to renewable energy. The Nyngan Solar Plant is part of the Australian Renewable Energy Agency's efforts to increase the country's solar capacity and reduce greenhouse gas emissions. The project has not only provided clean energy to thousands of homes but has also created jobs and stimulated economic growth in the region. Australia's commitment to solar energy is further demonstrated by its rapid adoption of rooftop solar panels, with millions of households generating their own electricity.

In Africa, the Noor Ouarzazate Solar Complex in Morocco is a shining example of solar energy's potential to transform energy systems in developing regions. Situated on the edge of the Sahara Desert, this solar complex is one of the largest concentrated solar power (CSP) plants in the world. It utilizes mirrors to focus sunlight onto a central tower, generating heat that is used to produce electricity. The Noor complex has a capacity of 580 megawatts and is part of Morocco's plan to generate 52% of its electricity from renewable sources by 2030. This project not only provides clean energy to millions of people but also positions Morocco as a leader in renewable energy in Africa.

Chile's Atacama Desert, known for its high solar irradiance, is home to the Cerro Dominador Solar Thermal Plant. This innovative project combines photovoltaic and concentrated solar power technologies to provide a reliable and continuous energy supply. The plant's unique design allows it to store energy in the form of molten salt, which can be used to generate electricity even when the sun is not shining. This capability addresses the intermittency challenge of solar power and ensures a stable energy supply. The Cerro Dominador project is a key component of Chile's strategy to diversify its energy mix and reduce its dependence on fossil fuels.

In the Middle East, the Mohammed bin Rashid Al Maktoum Solar Park in Dubai is a testament to the region's commitment to renewable energy. This solar park is one of the largest in the world, with a planned capacity of 5,000 megawatts by 2030. It features a combination of photovoltaic and concentrated solar power technologies, making it a versatile and efficient energy source. The solar park is part of Dubai's Clean Energy Strategy, which aims to generate 75% of the city's energy from renewable sources by 2050. The project's success is attributed to the region's abundant sunshine, supportive policies, and investment in cutting-edge technology.

These case studies illustrate the diverse ways in which solar energy is being harnessed around the world. From large-scale solar farms to innovative hybrid systems, these projects demonstrate the potential of solar power to provide clean, sustainable energy to millions of people. They also highlight the importance of supportive policies, technological innovation, and international collaboration in driving the global transition to renewable energy. As more countries embrace solar power, these successful projects serve as models for future development, paving the way for a brighter, more sustainable future.

Future Trends in Solar Energy Development

The future of solar energy development is poised to be transformative, driven by a confluence of technological advancements, policy initiatives, and market dynamics. As the world grapples with the urgent need to transition to sustainable energy sources, solar power stands at the forefront of this shift, offering a clean, abundant, and increasingly cost-effective solution.

One of the most promising trends in solar energy development is the continued improvement in photovoltaic (PV) cell efficiency. Researchers are constantly pushing the boundaries of what is possible, exploring new materials and technologies to enhance the performance of solar cells. Perovskite solar cells, for instance, have emerged as a game-changer in the field. These cells are made from a unique crystal structure that allows for high efficiency and low production costs. While still in the experimental phase, perovskite cells have the potential to surpass traditional silicon-based cells in terms of efficiency, making solar energy even more competitive with conventional energy sources.

In addition to advancements in cell efficiency, the integration of solar technology into everyday materials is gaining traction. Building-integrated photovoltaics (BIPV) represent a seamless way to incorporate solar energy into the architecture of buildings. By replacing conventional building materials with solar panels, BIPV systems allow structures to generate their own electricity without compromising aesthetics. This approach is particularly appealing for urban environments where space is limited, as it transforms rooftops, facades, and windows into energy-generating surfaces. As the cost of BIPV technology continues to decline, its adoption is expected to increase, contributing to the decentralization of energy production.

Energy storage solutions are also set to play a crucial role in the future of solar energy development. The intermittent nature of solar power—dependent on weather and daylight—poses challenges for grid stability and reliability. To address this, advancements in battery technology are essential. Lithium-ion batteries, which have become the standard for energy storage, are being continuously improved to increase their capacity, lifespan, and safety. Meanwhile, alternative storage technologies, such as flow batteries and solid-state batteries, are being explored for their potential to offer even greater performance. As storage solutions become more efficient and affordable, they will enable the widespread integration of solar energy into the grid, ensuring a steady and reliable power supply.

The rise of smart grid technology is another trend that is set to revolutionize solar energy development. Smart grids use digital technology to monitor and manage the flow of electricity, allowing for real-time adjustments to supply and demand. This capability is particularly important for integrating solar power, which can fluctuate based on environmental conditions. By optimizing energy distribution and reducing transmission losses, smart grids enhance the efficiency and reliability of solar energy systems. Furthermore, they empower consumers to take control of their energy use, enabling practices such as demand response and peer-to-peer energy trading.

Policy and regulatory frameworks will continue to shape the future of solar energy development. Governments around the world are implementing measures to encourage the adoption of solar power, including tax credits, subsidies, and renewable energy mandates. These policies not only make solar energy more accessible but also drive investment in research and development. As international agreements, such as the Paris Agreement, set ambitious targets for reducing carbon emissions, the pressure to transition to renewable energy sources will intensify, further accelerating the growth of solar energy.

The global solar market is also experiencing significant shifts, with emerging economies playing an increasingly important role. Countries in Asia, Africa, and Latin America are investing heavily in solar infrastructure to meet their growing energy demands and reduce their reliance on fossil fuels. These regions offer vast potential for solar energy development due to their abundant sunlight and expanding energy needs. As technology costs continue to decline, solar power is becoming an attractive option for these markets, driving global growth and diversification.

Innovation in solar panel design is another area of focus for future development. Researchers are exploring new ways to enhance the performance and versatility of solar panels, such as bifacial panels that capture sunlight on both sides and flexible panels that can be integrated into a variety of surfaces. These innovations not only increase energy output but also expand the range of applications for solar technology. As solar panels become more adaptable and efficient, they will open up new possibilities for harnessing solar energy in diverse environments.

The role of artificial intelligence and data analytics in solar energy development is also gaining attention. By analyzing vast amounts of data, AI can optimize the performance of solar systems, predict maintenance needs, and improve energy forecasting. This capability enhances the efficiency and reliability of solar installations, making them more attractive to consumers and investors alike. As AI technology continues to evolve, its integration into solar energy systems is expected to drive further advancements and efficiencies.

In conclusion, the future of solar energy development is bright, with numerous trends converging to drive its growth and adoption. From technological innovations and policy support to market dynamics and global collaboration, solar power is poised to play an increasingly central role in the transition to a sustainable energy future. As these trends continue to unfold, the potential for solar energy to transform the global energy landscape becomes ever more tangible, offering a path to a cleaner, more resilient world.

Chapter 2: Wind Power: Capturing the Breeze

The Science Behind Wind Energy and Its Impact

Harnessing the power of the wind has been a pursuit of humanity for centuries, from the simple windmills of the past to the sophisticated wind turbines of today. The science behind wind energy is both fascinating and complex, involving principles of physics, meteorology, and engineering. Understanding these principles is key to appreciating how wind energy is captured and converted into electricity, as well as its broader impact on the environment and society.

At its core, wind energy is derived from the movement of air masses in the Earth's atmosphere. This movement is primarily driven by the uneven heating of the Earth's surface by the sun. As sunlight warms the air, it rises, creating areas of low pressure. Cooler air then moves in to replace the rising warm air, generating wind. The kinetic energy of this moving air can be harnessed and converted into mechanical energy using wind turbines.

Wind turbines are the workhorses of wind energy production. These towering structures consist of several key components: the rotor blades, the nacelle, and the tower. The rotor blades are designed to capture the wind's kinetic energy. As the wind blows, it causes the blades to rotate, much like the sails of a windmill. This rotation turns a shaft connected to a generator housed within the nacelle, converting the mechanical energy into electrical energy. The tower elevates the rotor blades to capture stronger and more consistent winds found at higher altitudes.

The efficiency of a wind turbine is influenced by several factors, including the design of the blades, the height of the tower, and the wind speed. Modern turbines are equipped with aerodynamic blades that are optimized to capture the maximum amount of energy from the wind. The height of the tower is also crucial, as wind speeds generally increase with altitude. By placing turbines on taller towers, they can access more powerful winds, increasing their energy output.

Wind speed is a critical factor in wind energy production. The power generated by a wind turbine is proportional to the cube of the wind speed, meaning that even small increases in wind speed can lead to significant increases in energy output. This relationship underscores the importance of site selection for wind farms. Ideal locations are those with consistent and strong winds, such as coastal areas, open plains, and hilltops. Meteorological studies and wind resource assessments are conducted to identify these optimal sites.

The impact of wind energy extends beyond its ability to generate electricity. As a renewable energy source, wind power offers significant environmental benefits. Unlike fossil fuels, wind energy production does not emit greenhouse gases or other pollutants, making it a clean and sustainable alternative. By reducing reliance on fossil fuels, wind energy helps mitigate climate change and improve air quality, contributing to a healthier environment.

Wind energy also has economic implications. The development of wind farms creates jobs in manufacturing, installation, and maintenance, stimulating local economies. Additionally, wind energy can provide a stable and predictable source of income for landowners who lease their land for turbine installation. As technology advances and economies of scale are realized, the cost of wind energy continues to decline, making it increasingly competitive with traditional energy sources.

Despite its many advantages, wind energy is not without challenges. One of the primary concerns is the intermittency of wind, as it is not always blowing when electricity is needed. This variability can pose challenges for grid stability and reliability. To address this, advancements in energy storage technologies, such as batteries and pumped hydro storage, are being developed to store excess energy generated during periods of high wind for use when the wind is calm.

Another challenge is the potential impact of wind turbines on wildlife, particularly birds and bats. While the overall impact is relatively low compared to other human activities, it is important to carefully assess and mitigate these effects through thoughtful site selection and turbine design. Research and monitoring efforts are ongoing to better understand and minimize the impact on wildlife.

Public perception and acceptance of wind energy can also influence its development. While many people support the use of renewable energy, concerns about noise, visual impact, and land use can lead to opposition from local communities. Engaging with stakeholders, providing transparent information, and addressing concerns are essential to building support for wind energy projects.

The future of wind energy is promising, with ongoing research and innovation driving further advancements. Floating wind turbines, for example, offer the potential to harness wind energy in deep offshore waters, where winds are often stronger and more consistent. This technology could significantly expand the areas available for wind energy development, increasing its contribution to the global energy mix.

In conclusion, the science behind wind energy is a testament to human ingenuity and our ability to harness the natural forces of the Earth for sustainable power. By understanding the principles and challenges of wind energy, we can continue to innovate and expand its use, paving the way for a cleaner and more sustainable energy future. As we navigate the complexities of energy transition, wind power stands as a vital component of our efforts to create a resilient and environmentally responsible energy system.

Onshore vs. Offshore Wind Farms: A Comparative Analysis

Wind energy has become a cornerstone of the global transition to renewable energy, with both onshore and offshore wind farms playing pivotal roles. Each type of wind farm offers unique advantages and faces distinct challenges, making it essential to understand their differences to make informed decisions about their development and deployment.

Onshore wind farms are the most common type of wind energy installation. They are typically located in rural areas, open plains, or hilltops where wind speeds are favorable. One of the primary advantages of onshore wind farms is their relatively lower cost of installation and maintenance compared to offshore counterparts. The infrastructure required for onshore wind farms, such as roads and electrical connections, is generally less complex and expensive to build. Additionally, onshore wind farms benefit from easier access for maintenance and repairs, reducing operational costs and downtime.

The proximity of onshore wind farms to existing power grids is another significant advantage. This closeness allows for more straightforward integration of the generated electricity into the grid, minimizing transmission losses and reducing the need for extensive infrastructure upgrades. Onshore wind farms also tend to have shorter development timelines, as the permitting and construction processes are typically less complicated than those for offshore projects.

However, onshore wind farms are not without their challenges. One of the most significant issues is the availability of suitable land. As demand for wind energy grows, finding locations with optimal wind conditions that do not conflict with other land uses, such as agriculture or conservation, becomes increasingly difficult. Additionally, onshore wind farms can face opposition from local communities due to concerns about noise, visual impact, and potential effects on property values.

Offshore wind farms, on the other hand, are located in bodies of water, usually on the continental shelf where the water is relatively shallow. One of the most compelling advantages of offshore wind farms is the availability of stronger and more consistent winds compared to onshore sites. This results in higher energy output and greater efficiency, making offshore wind farms an attractive option for maximizing energy production.

The vast expanses of open water also provide ample space for large-scale wind farm development, reducing competition for land and minimizing conflicts with other land uses. Offshore wind farms can be situated far from populated areas, mitigating concerns about noise and visual impact. This distance from shore also allows for the construction of larger turbines, which can capture more energy and further increase efficiency.

Despite these advantages, offshore wind farms face several challenges that can impact their feasibility and cost-effectiveness. The harsh marine environment presents significant technical and logistical challenges, from the construction and installation of turbines to their ongoing maintenance. Saltwater corrosion, strong currents, and extreme weather conditions can all affect the durability and reliability of offshore wind installations.

The cost of offshore wind farms is generally higher than that of onshore projects, primarily due to the complexities of construction and maintenance in a marine setting. Specialized vessels and equipment are required for installation, and the logistics of transporting materials and personnel to and from the site can be costly and time-consuming. Additionally, the integration of offshore wind energy into the power grid requires the construction of undersea cables and substations, further increasing costs.

Environmental considerations are also a critical factor in the development of both onshore and offshore wind farms. Onshore wind farms must carefully assess their impact on local ecosystems, including potential effects on wildlife and habitats. Offshore wind farms, while avoiding some of the land-based environmental concerns, must consider their impact on marine life and ecosystems. Careful site selection, environmental assessments, and mitigation measures are essential to minimize these impacts and ensure sustainable development.

The choice between onshore and offshore wind farms ultimately depends on a variety of factors, including geographic location, wind resources, environmental considerations, and economic feasibility. In some cases, a combination of both onshore and offshore wind energy may be the most effective strategy for meeting energy needs and achieving renewable energy targets.

As technology continues to advance, the potential for both onshore and offshore wind farms to contribute to the global energy mix will only increase. Innovations in turbine design, materials, and installation techniques are expected to reduce costs and improve the efficiency and reliability of wind energy systems. Floating wind turbines, for example, offer the potential to expand offshore wind energy into deeper waters, where traditional fixed-bottom turbines are not feasible.

In conclusion, both onshore and offshore wind farms offer valuable contributions to the renewable energy landscape, each with its own set of advantages and challenges. By understanding the differences between these two types of wind energy installations, stakeholders can make informed decisions about their development and deployment, ensuring that wind energy continues to play a vital role in the transition to a sustainable energy future. As we navigate the complexities of energy transition, the complementary strengths of onshore and offshore wind farms provide a robust foundation for a cleaner, more resilient energy system.

Policy and Regulation Supporting Wind Energy Growth

The growth of wind energy as a pivotal component of the global renewable energy landscape is inextricably linked to the policies and regulations that support its development. Governments around the world have recognized the potential of wind power to reduce carbon emissions, enhance energy security, and stimulate economic growth. As a result, a variety of policy instruments and regulatory frameworks have been implemented to encourage investment in wind energy and facilitate its integration into the energy mix.

One of the most effective policy tools for promoting wind energy is the establishment of renewable energy targets. These targets set specific goals for the proportion of energy that must be generated from renewable sources, including wind, within a given timeframe. By providing a clear and measurable objective, renewable energy targets create a stable and predictable market environment that encourages investment in wind energy projects. Countries such as Germany, Denmark, and China have set ambitious renewable energy targets, driving significant growth in their wind energy sectors.

Feed-in tariffs (FITs) have been instrumental in supporting the expansion of wind energy. These tariffs guarantee a fixed price for the electricity generated by wind farms over a specified period, providing investors with a reliable revenue stream. By reducing financial risk, FITs have attracted investment in wind energy projects and accelerated their deployment. While some countries have phased out FITs in favor of other mechanisms, they remain a valuable tool in regions where wind energy is still emerging.

Renewable portfolio standards (RPS) are another policy mechanism that has been used to promote wind energy. RPS mandates that a certain percentage of electricity sold by utilities must come from renewable sources. This requirement creates a market for renewable energy certificates (RECs), which can be traded to meet compliance obligations. By incentivizing utilities to invest in wind energy, RPS policies have driven the growth of wind power in countries such as the United States and Australia.

Tax incentives and subsidies also play a crucial role in supporting wind energy development. Investment tax credits (ITCs) and production tax credits (PTCs) provide financial benefits to wind energy developers, reducing the cost of project development and operation. These incentives have been particularly effective in the United States, where they have contributed to the rapid expansion of the wind energy sector. Additionally, grants and low-interest loans can help offset the high upfront costs associated with wind energy projects, making them more accessible to developers.

Regulatory frameworks are essential for ensuring the smooth integration of wind energy into the power grid. Grid connection standards and procedures must be established to facilitate the connection of wind farms to the electricity network. These standards ensure that wind energy can be reliably and efficiently transmitted to consumers, minimizing disruptions and maintaining grid stability. In some regions, grid operators have implemented priority dispatch rules, which give preference to renewable energy sources, including wind, over conventional power plants.

Environmental regulations are also a critical consideration in the development of wind energy projects. While wind power is a clean and sustainable energy source, its development can have environmental impacts, such as effects on wildlife and habitats.

Environmental impact assessments (EIAs) are required to evaluate these potential impacts and identify mitigation measures. By ensuring that wind energy projects are developed in an environmentally responsible manner, regulations help balance the need for renewable energy with the protection of natural resources.

Public engagement and community involvement are important aspects of policy and regulation supporting wind energy growth. Transparent communication and consultation with local communities can help address concerns about wind energy projects, such as noise, visual impact, and land use. By involving stakeholders in the decision-making process, policymakers can build public support for wind energy and facilitate its development.

International cooperation and collaboration are also vital for advancing wind energy. Organizations such as the International Renewable Energy Agency (IRENA) and the Global Wind Energy Council (GWEC) provide platforms for sharing knowledge, best practices, and policy experiences. By working together, countries can accelerate the deployment of wind energy and overcome common challenges, such as technology transfer and capacity building.

As the wind energy sector continues to evolve, policy and regulation must adapt to changing circumstances and emerging technologies. Innovations such as floating wind turbines and hybrid energy systems present new opportunities and challenges for policymakers. By staying informed and responsive to these developments, governments can ensure that their policies remain effective and supportive of wind energy growth.

In conclusion, the success of wind energy as a key component of the global energy transition is heavily dependent on the policies and regulations that support its development. By implementing a combination of renewable energy targets, financial incentives, regulatory frameworks, and public engagement strategies, governments can create an enabling environment for wind energy growth. As the world moves towards a more sustainable energy future, the continued support of wind energy through effective policy and regulation will be essential for achieving climate goals and ensuring energy security.

Community and Environmental Considerations

When embarking on the journey of developing wind energy projects, the interplay between community interests and environmental considerations becomes a crucial aspect of the planning and implementation process. These projects, while offering significant benefits in terms of renewable energy generation and reduced carbon emissions, must be carefully balanced with the needs and concerns of local communities and the natural environment.

Community engagement is a cornerstone of successful wind energy development. The presence of wind farms can evoke a range of responses from local residents, from enthusiastic support to staunch opposition. Understanding and addressing these perspectives is essential for fostering a positive relationship between developers and the community. Open and transparent communication is key. By involving community members early in the planning process, developers can build trust and address concerns related to noise, visual impact, and land use. Public meetings, workshops, and informational sessions provide platforms for dialogue, allowing residents to voice their opinions and learn about the benefits and potential impacts of the project.

One of the primary concerns for communities is the visual impact of wind turbines on the landscape. The towering structures can alter the aesthetic character of an area, leading to resistance from those who value the natural beauty of their surroundings. To mitigate these concerns, developers can explore design options that minimize visual intrusion, such as selecting turbine colors that blend with the environment or strategically placing turbines to reduce their prominence. Additionally, engaging with local artists and architects can lead to creative solutions that integrate turbines into the landscape in a harmonious manner.

Noise is another common concern associated with wind farms. The sound generated by the rotation of turbine blades can be a source of annoyance for nearby residents. To address this issue, developers can conduct thorough noise assessments and implement measures to minimize sound levels, such as selecting low-noise turbine models and maintaining appropriate distances from residential areas. By proactively addressing noise concerns, developers can alleviate potential conflicts and enhance community acceptance.

Economic benefits are a compelling aspect of wind energy projects that can garner community support. Wind farms can create jobs during the construction and operational phases, providing employment opportunities for local residents. Additionally, landowners who lease their land for turbine installation can receive a steady income stream, contributing to the local economy. Developers can further enhance community benefits by establishing community funds or offering shares in the project, allowing residents to directly benefit from the success of the wind farm.

Environmental considerations are equally important in the development of wind energy projects. While wind power is a clean and sustainable energy source, its implementation can have ecological impacts that must be carefully managed. One of the primary environmental concerns is the potential impact on wildlife, particularly birds and bats. Wind turbines can pose a collision risk for these animals, leading to fatalities. To mitigate this risk, developers can conduct comprehensive environmental impact assessments to identify sensitive areas and implement measures such as turbine curtailment during peak migration periods or the use of technology to detect and deter wildlife.

The siting of wind farms is a critical factor in minimizing environmental impacts. By selecting locations that avoid ecologically sensitive areas, such as wetlands, forests, and habitats of endangered species, developers can reduce the potential for negative effects on biodiversity. Collaboration with environmental organizations and wildlife experts can provide valuable insights into site selection and impact mitigation strategies.

The construction and operation of wind farms can also affect local ecosystems through habitat disturbance and land use changes. To minimize these impacts, developers can implement best practices for construction, such as minimizing land clearing, restoring disturbed areas, and using environmentally friendly materials. Ongoing monitoring and adaptive management strategies can ensure that any unforeseen impacts are promptly addressed, maintaining the ecological integrity of the area.

Community and environmental considerations are not mutually exclusive; rather, they are interconnected aspects of wind energy development that must be addressed in tandem. By adopting a holistic approach that prioritizes both community engagement and environmental stewardship, developers can create projects that are not only technically and economically viable but also socially and ecologically responsible.

The role of policy and regulation in supporting community and environmental considerations cannot be overstated. Governments can establish guidelines and standards that promote best practices in community engagement and environmental protection. Incentives for developers who demonstrate exemplary performance in these areas can further encourage responsible development. Additionally, regulatory frameworks can facilitate collaboration between developers, communities, and environmental organizations, fostering a cooperative approach to wind energy development.

In conclusion, the successful development of wind energy projects hinges on the careful consideration of community and environmental factors. By engaging with local residents, addressing their concerns, and implementing measures to protect the natural environment, developers can create projects that are embraced by communities and contribute positively to the global transition to renewable energy. As the demand for clean energy continues to grow, the integration of community and environmental considerations will remain a vital component of sustainable wind energy development.

Innovations and Future Directions in Wind Technology

The evolution of wind technology has been marked by remarkable innovations that have transformed it into one of the most promising sources of renewable energy. As the world continues to seek sustainable solutions to meet growing energy demands, the future of wind technology is poised for even greater advancements. These innovations not only aim to enhance the efficiency and cost-effectiveness of wind energy but also to expand its applicability and integration into the global energy landscape.

One of the most significant areas of innovation in wind technology is the development of larger and more efficient wind turbines. The trend towards larger turbines is driven by the desire to capture more energy from the wind, thereby increasing the overall output of wind farms. Modern turbines are now reaching heights of over 200 meters, with rotor diameters exceeding 150 meters. These colossal structures are capable of generating significantly more electricity than their predecessors, making them ideal for both onshore and offshore applications. The use of advanced materials, such as carbon fiber and lightweight composites, has enabled the construction of these larger turbines without compromising structural integrity.

The design of turbine blades is also undergoing a transformation. Engineers are exploring innovative blade shapes and configurations to optimize aerodynamic performance and reduce noise. One promising development is the use of biomimicry, where blade designs are inspired by the natural world. For example, the serrated edges of humpback whale fins have inspired the design of turbine blades with similar features, which can improve lift and reduce drag. These bio-inspired designs have the potential to enhance the efficiency of wind turbines, particularly in low-wind conditions.

Floating wind turbines represent a groundbreaking innovation that is set to revolutionize offshore wind energy. Traditional offshore turbines are anchored to the seabed, limiting their deployment to shallow waters. Floating turbines, however, are mounted on buoyant platforms that can be anchored in deeper waters, where wind speeds are typically higher and more consistent. This technology opens up vast new areas for wind energy development, particularly in regions with deep coastal waters. Several pilot projects are already underway, demonstrating the feasibility and potential of floating wind farms to contribute significantly to the global energy mix.

The integration of digital technologies and data analytics is another frontier in wind technology innovation. The use of sensors and real-time data collection allows for the continuous monitoring of turbine performance and environmental conditions. This data can be analyzed using advanced algorithms to optimize turbine operation, predict maintenance needs, and improve energy forecasting. By harnessing the power of data, operators can enhance the efficiency and reliability of wind farms, reducing downtime and operational costs.

Energy storage solutions are becoming increasingly important as wind energy continues to grow. The intermittent nature of wind power necessitates the development of effective storage systems to ensure a stable and reliable energy supply. Innovations in battery technology, such as lithium-ion and flow batteries, are being explored to store excess energy generated during periods of high wind. Additionally, hybrid systems that combine wind energy with other renewable sources, such as solar or hydropower, offer a promising approach to balancing supply and demand.

The concept of wind-solar hybrid systems is gaining traction as a means of maximizing renewable energy generation. By co-locating wind turbines and solar panels, these systems can take advantage of complementary energy sources, as wind and solar generation often peak at different times. This synergy can lead to more consistent energy output and improved utilization of available land and infrastructure. Hybrid systems also offer the potential for shared grid connections and maintenance resources, further enhancing their economic viability.

The future of wind technology is also being shaped by advancements in grid integration and smart grid solutions. As wind energy becomes a larger part of the energy mix, the ability to efficiently integrate it into the power grid is crucial. Smart grids use digital technology to monitor and manage the flow of electricity, allowing for real-time adjustments to supply and demand. This capability is particularly important for accommodating the variability of wind power and ensuring grid stability. By optimizing energy distribution and reducing transmission losses, smart grids enhance the overall efficiency and reliability of wind energy systems.

Policy and regulatory frameworks will continue to play a vital role in supporting the development and deployment of innovative wind technologies. Governments can incentivize research and development through grants, subsidies, and tax credits, encouraging investment in cutting-edge solutions. International collaboration and knowledge sharing can also accelerate the adoption of best practices and technological advancements, driving the global expansion of wind energy.

As we look to the future, the potential for wind technology to transform the global energy landscape is immense. Continued innovation and investment in research and development will be essential for overcoming existing challenges and unlocking new opportunities. By embracing these advancements, we can harness the full potential of wind energy to create a cleaner, more sustainable energy future for generations to come. The journey of wind technology is far from over, and its future promises to be as dynamic and transformative as its past.

Chapter 3: Hydropower: The Flow of Change

Basics of Hydroelectric Power Generation

Hydroelectric power generation stands as one of the oldest and most reliable forms of renewable energy, harnessing the natural flow of water to produce electricity. Its roots trace back to ancient civilizations that utilized water wheels for mechanical tasks, but it was not until the late 19th century that hydroelectric power began to be used for electricity generation. Today, it accounts for a significant portion of the world's renewable energy supply, offering a clean and sustainable alternative to fossil fuels.

At the heart of hydroelectric power generation is the conversion of potential energy stored in water into mechanical energy, and subsequently into electrical energy. This process typically involves the construction of a dam on a river, creating a reservoir or a large body of water. The dam serves as a barrier that controls the flow of water, allowing it to be released in a controlled manner. As water is released from the reservoir, it flows through turbines, causing them to spin. These turbines are connected to generators, which convert the mechanical energy of the spinning turbines into electrical energy.

The efficiency of hydroelectric power generation is largely dependent on the height from which the water falls, known as the "head," and the volume of water flow. A higher head and greater water flow result in more energy being generated. This principle is why hydroelectric plants are often located in mountainous regions or areas with significant elevation changes, where the natural topography provides the necessary conditions for efficient energy production.

There are several types of hydroelectric power plants, each suited to different geographical and environmental conditions. The most common type is the impoundment facility, which uses a dam to store river water in a reservoir. This stored water can be released as needed to generate electricity, providing a reliable and flexible energy source. Another type is the run-of-the-river facility, which does not require a large reservoir. Instead, it diverts a portion of the river's flow through a channel or penstock to generate electricity. This type of plant has a smaller environmental footprint but is more dependent on river flow conditions.

Pumped storage is a unique form of hydroelectric power that acts as a large-scale energy storage system. During periods of low electricity demand, excess energy is used to pump water from a lower reservoir to an upper reservoir. When demand increases, the stored water is released back to the lower reservoir, passing through turbines to generate electricity. This system provides a valuable means of balancing supply and demand, particularly in grids with a high penetration of intermittent renewable energy sources like wind and solar.

The environmental impact of hydroelectric power generation is a topic of considerable discussion. While it is a clean energy source that does not produce direct emissions, the construction and operation of hydroelectric plants can have significant ecological effects. The creation of reservoirs can lead to the flooding of large areas, impacting local ecosystems and displacing communities. Changes in water flow can affect aquatic habitats and fish populations, necessitating careful management and mitigation strategies.

Fish ladders and bypass systems are examples of measures that can be implemented to minimize the impact on aquatic life. These structures allow fish to navigate around dams and continue their natural migration patterns. Additionally, environmental flow requirements can be established to ensure that sufficient water is released downstream to maintain the health of river ecosystems.

The social implications of hydroelectric power generation are also important to consider. While these projects can provide substantial economic benefits, including job creation and increased energy security, they can also lead to the displacement of communities and changes in land use. Engaging with local stakeholders and ensuring that affected communities are involved in the decision-making process is crucial for the equitable development of hydroelectric projects.

Technological advancements continue to enhance the efficiency and sustainability of hydroelectric power generation. Innovations in turbine design, such as the development of fish-friendly turbines, aim to reduce environmental impacts while maintaining high levels of energy production. The integration of digital technologies and data analytics allows for more precise monitoring and control of hydroelectric plants, optimizing performance and reducing maintenance costs.

Small-scale hydroelectric systems, often referred to as micro-hydro or mini-hydro, offer opportunities for decentralized energy generation, particularly in remote or rural areas. These systems can provide reliable and affordable electricity to communities that are not connected to the main power grid, supporting local development and improving quality of life.

The future of hydroelectric power generation is promising, with the potential to play a key role in the transition to a sustainable energy future. By balancing the need for clean energy with environmental and social considerations, hydroelectric power can continue to provide a reliable and renewable source of electricity for generations to come. As we strive to meet global energy demands while reducing our carbon footprint, the continued development and innovation in hydroelectric technology will be essential in achieving these goals.

The Role of Small-Scale Hydropower in the Energy Shift

Small-scale hydropower, often referred to as micro-hydro or mini-hydro, is emerging as a vital component in the global shift towards sustainable energy. Unlike large-scale hydroelectric projects, which can have significant environmental and social impacts, small-scale hydropower offers a more localized and environmentally friendly approach to energy generation. Its role in the energy transition is becoming increasingly important, particularly in remote and rural areas where access to reliable electricity is limited.

The appeal of small-scale hydropower lies in its ability to harness the natural flow of water without the need for large dams or reservoirs. These systems typically generate up to 10 megawatts of electricity, making them suitable for powering small communities, individual homes, or even specific industrial applications. By utilizing existing watercourses, such as rivers, streams, or irrigation canals, small-scale hydropower can provide a consistent and renewable source of energy with minimal disruption to the environment.

One of the key advantages of small-scale hydropower is its potential for decentralized energy generation. In many parts of the world, particularly in developing countries, centralized power grids are either non-existent or unreliable. Small-scale hydropower systems can be installed in remote locations, providing electricity to communities that are not connected to the main grid. This decentralization not only enhances energy security but also empowers local communities by giving them control over their energy resources.

The installation of small-scale hydropower systems can have a transformative impact on rural communities. Access to electricity can improve quality of life by enabling the use of modern appliances, lighting, and communication technologies. It can also support local economic development by powering small businesses, agricultural operations, and educational facilities. In many cases, the availability of reliable electricity can lead to improved health outcomes, as communities gain access to clean water, refrigeration for medicines, and better healthcare facilities.

Environmental sustainability is a significant consideration in the development of small-scale hydropower. These systems are designed to have a minimal ecological footprint, often requiring little to no alteration of the natural watercourse. By avoiding the construction of large dams, small-scale hydropower reduces the risk of habitat destruction, fish migration disruption, and changes in water quality. Additionally, the use of run-of-the-river designs ensures that water flow is maintained, preserving the ecological balance of the river system.

The implementation of small-scale hydropower projects requires careful planning and consideration of local conditions. Site selection is critical, as the availability of a suitable water source with sufficient flow and head is essential for efficient energy generation. Hydrological assessments and feasibility studies are necessary to determine the potential energy output and ensure the sustainability of the project. Engaging with local communities and stakeholders is also crucial to address any concerns and ensure that the benefits of the project are shared equitably.

Technological advancements are playing a significant role in enhancing the efficiency and accessibility of small-scale hydropower. Innovations in turbine design, such as the development of low-head and ultra-low-head turbines, are expanding the range of sites that can be utilized for hydropower generation. These turbines are capable of operating efficiently in locations with minimal elevation changes, making them ideal for flat or gently sloping terrains. Additionally, the use of modular and prefabricated components can reduce installation costs and simplify the construction process.

The integration of small-scale hydropower with other renewable energy sources offers exciting opportunities for hybrid energy systems. By combining hydropower with solar, wind, or biomass energy, communities can create a more resilient and reliable energy supply. These hybrid systems can take advantage of the complementary nature of different energy sources, ensuring a consistent power supply even when one source is unavailable. For example, solar panels can provide electricity during the day, while hydropower can supply energy at night or during cloudy periods.

Policy and regulatory support are essential for the successful deployment of small-scale hydropower. Governments can encourage investment in these projects by providing financial incentives, such as grants, subsidies, or tax credits. Streamlined permitting processes and clear regulatory frameworks can also facilitate the development of small-scale hydropower, reducing administrative barriers and encouraging innovation. International cooperation and knowledge sharing can further accelerate the adoption of best practices and technological advancements.

The role of small-scale hydropower in the energy shift is multifaceted, offering a sustainable and adaptable solution to the challenges of energy access and environmental conservation. By harnessing the power of water in a responsible and efficient manner, small-scale hydropower can contribute to a cleaner, more equitable energy future. As the world continues to transition towards renewable energy, the potential of small-scale hydropower to empower communities and support sustainable development will remain a vital component of the global energy landscape.

Environmental and Social Impacts of Hydropower Projects

Hydropower projects, while offering a renewable source of energy, present a complex array of environmental and social impacts that must be carefully managed to ensure sustainable development. The construction and operation of these projects can lead to significant changes in ecosystems and communities, necessitating a balanced approach that considers both the benefits and the potential drawbacks.

The environmental impacts of hydropower projects are primarily associated with the alteration of natural watercourses and the creation of reservoirs. One of the most significant ecological consequences is the disruption of aquatic habitats. Dams and reservoirs can impede the natural migration patterns of fish and other aquatic species, leading to population declines and loss of biodiversity. To mitigate these effects, fish ladders and bypass systems can be installed to facilitate the movement of aquatic life around barriers. Additionally, environmental flow requirements can be established to ensure that sufficient water is released downstream to maintain the health of river ecosystems.

The inundation of land to create reservoirs can also have profound effects on terrestrial ecosystems. Large areas of forest, wetlands, and agricultural land may be submerged, resulting in habitat loss and fragmentation. This can lead to the displacement of wildlife and changes in local biodiversity. To address these impacts, comprehensive environmental impact assessments are conducted to identify sensitive areas and develop strategies for habitat restoration and conservation. Reforestation and the creation of protected areas can help offset the loss of natural habitats and support biodiversity conservation.

Water quality is another critical concern associated with hydropower projects. The alteration of natural water flow can lead to changes in sediment transport and nutrient cycling, affecting the physical and chemical properties of the water. Reservoirs can become stratified, with layers of water at different temperatures and oxygen levels, leading to the development of anoxic conditions that can harm aquatic life. To manage these issues, water quality monitoring programs are implemented to track changes and inform adaptive management strategies. Techniques such as selective withdrawal and aeration can be used to improve water quality and maintain healthy aquatic ecosystems.

The social impacts of hydropower projects are closely linked to the environmental changes they bring about. The creation of reservoirs often requires the relocation of communities, leading to the displacement of people and the loss of homes, livelihoods, and cultural heritage. This can result in social and economic disruption, particularly for indigenous and marginalized communities who may have strong cultural ties to the land. To address these challenges, comprehensive resettlement and compensation plans are developed in consultation with affected communities. These plans aim to provide fair compensation, support livelihood restoration, and ensure that displaced people have access to essential services and infrastructure.

The construction and operation of hydropower projects can also lead to changes in land use and access to natural resources. Communities that rely on fishing, agriculture, or forestry may experience reduced access to these resources, impacting their livelihoods and food security. Engaging with local stakeholders and incorporating their knowledge and perspectives into project planning is essential for identifying and addressing these impacts. Collaborative management approaches that involve communities in decision-making can help ensure that the benefits of hydropower projects are shared equitably and that negative impacts are minimized.

Despite these challenges, hydropower projects can also bring significant social and economic benefits. They provide a reliable source of clean energy, contributing to energy security and reducing reliance on fossil fuels. The construction and operation of hydropower facilities can create jobs and stimulate local economies, providing opportunities for skill development and capacity building. Additionally, the infrastructure associated with hydropower projects, such as roads and bridges, can improve access to remote areas and support regional development.

The role of policy and regulation is crucial in managing the environmental and social impacts of hydropower projects. Governments can establish guidelines and standards that promote best practices in environmental management and social responsibility. Regulatory frameworks can facilitate stakeholder engagement and ensure that the rights and interests of affected communities are respected. International cooperation and knowledge sharing can further enhance the capacity of countries to implement sustainable hydropower projects.

Technological advancements are also playing a role in reducing the environmental and social impacts of hydropower. Innovations in turbine design, such as fish-friendly turbines, aim to minimize harm to aquatic life while maintaining high levels of energy production. The use of digital technologies and data analytics allows for more precise monitoring and control of hydropower facilities, optimizing performance and reducing environmental impacts.

In conclusion, the development of hydropower projects requires a comprehensive understanding of their environmental and social impacts. By adopting a holistic approach that prioritizes sustainability and stakeholder engagement, it is possible to harness the benefits of hydropower while minimizing its negative effects. As the world continues to transition towards renewable energy, the careful management of hydropower projects will be essential in achieving a balance between energy needs and environmental conservation. Through collaboration, innovation, and responsible governance, hydropower can play a vital role in a sustainable energy future.

Advances in Marine and Hydrokinetic Energy

The vast expanse of the world's oceans and rivers holds immense potential for energy generation, a promise that is being increasingly realized through advances in marine and hydrokinetic energy technologies. These innovations are paving the way for harnessing the power of water in ways that are both sustainable and efficient, contributing to the diversification of the renewable energy portfolio and offering new opportunities for clean energy production.

Marine energy, often referred to as ocean energy, encompasses a variety of technologies designed to capture the energy from ocean waves, tides, and currents. One of the most promising areas of marine energy is wave energy, which exploits the kinetic and potential energy of ocean surface waves. Wave energy converters (WECs) are devices that capture this energy and convert it into electricity. These devices come in various forms, including point absorbers, oscillating water columns, and attenuators, each suited to different wave conditions and environments. Advances in materials science and engineering are enhancing the efficiency and durability of WECs, enabling them to withstand the harsh marine environment and operate effectively over long periods.

Tidal energy is another key component of marine energy, harnessing the gravitational pull of the moon and sun on the Earth's oceans to generate electricity. Tidal energy systems can be broadly categorized into tidal stream and tidal range technologies. Tidal stream systems capture the kinetic energy of moving water in tidal currents using underwater turbines, similar to wind turbines. These systems are particularly effective in areas with strong tidal flows, such as narrow straits and coastal inlets. Tidal range systems, on the other hand, utilize the potential energy created by the difference in height between high and low tides. This is typically achieved through the construction of barrages or lagoons that capture and release water to drive turbines. Recent innovations in turbine design and energy storage are improving the efficiency and viability of tidal energy systems, making them a more attractive option for renewable energy generation.

Hydrokinetic energy, which refers to the energy derived from the movement of water in rivers and streams, is another area of significant advancement. Unlike traditional hydropower, which often requires large dams and reservoirs, hydrokinetic systems can be deployed in free-flowing water bodies with minimal environmental impact. These systems use underwater turbines to capture the kinetic energy of flowing water, converting it into electricity. Advances in turbine technology, such as the development of low-head and ultra-low-head turbines, are expanding the range of sites suitable for hydrokinetic energy generation. These innovations are enabling the deployment of hydrokinetic systems in areas with gentle slopes and low water velocities, increasing the accessibility and applicability of this technology.

The integration of digital technologies and data analytics is playing a crucial role in the advancement of marine and hydrokinetic energy. Real-time monitoring and control systems allow for the optimization of energy capture and conversion, enhancing the efficiency and reliability of these technologies. Data-driven insights can inform maintenance schedules and operational strategies, reducing downtime and extending the lifespan of energy systems. Additionally, the use of predictive modeling and simulation tools is aiding in the design and testing of new technologies, accelerating the development and deployment of innovative solutions.

Environmental considerations are a critical aspect of marine and hydrokinetic energy development. The deployment of these technologies must be carefully managed to minimize impacts on marine and freshwater ecosystems. Potential environmental effects include changes in water flow, habitat disruption, and impacts on aquatic life. To address these concerns, comprehensive environmental impact assessments are conducted to identify and mitigate potential risks. The development of environmentally friendly technologies, such as fish-friendly turbines and low-impact anchoring systems, is helping to reduce the ecological footprint of marine and hydrokinetic energy projects.

The social and economic benefits of marine and hydrokinetic energy are significant. These technologies have the potential to provide a reliable and sustainable source of electricity to coastal and riverine communities, enhancing energy security and supporting local development. The construction and operation of marine and hydrokinetic energy projects can create jobs and stimulate economic growth, providing opportunities for skill development and capacity building. Additionally, the diversification of the energy mix through the inclusion of marine and hydrokinetic energy can reduce reliance on fossil fuels, contributing to the reduction of greenhouse gas emissions and the mitigation of climate change.

Policy and regulatory frameworks play a vital role in supporting the development and deployment of marine and hydrokinetic energy. Governments can encourage investment in these technologies through financial incentives, such as grants, subsidies, and tax credits. Clear regulatory guidelines and streamlined permitting processes can facilitate the development of projects, reducing administrative barriers and encouraging innovation. International collaboration and knowledge sharing can further accelerate the adoption of best practices and technological advancements, driving the global expansion of marine and hydrokinetic energy.

The future of marine and hydrokinetic energy is promising, with the potential to play a key role in the transition to a sustainable energy future. Continued innovation and investment in research and development will be essential for overcoming existing challenges and unlocking new opportunities. By embracing these advancements, we can harness the full potential of marine and hydrokinetic energy to create a cleaner, more sustainable energy future for generations to come. As the demand for renewable energy continues to grow, the integration of marine and hydrokinetic energy into the global energy landscape will remain a vital component of the transition to a low-carbon economy.

The Future of Hydropower in a Sustainable Energy Mix

The future of hydropower in a sustainable energy mix is a topic of growing importance as the world seeks to transition to cleaner and more reliable sources of energy. Hydropower, with its long history and proven track record, is poised to play a significant role in this transition. However, its future will be shaped by a combination of technological advancements, environmental considerations, and evolving energy policies.

Hydropower's potential lies in its ability to provide a stable and continuous source of electricity, which is crucial for balancing the intermittent nature of other renewable energy sources like wind and solar. As the demand for renewable energy increases, hydropower can serve as a backbone for the energy grid, offering both base-load power and the flexibility to ramp up or down in response to fluctuations in demand. This capability makes it an ideal partner for integrating variable renewable energy sources, ensuring a reliable and resilient energy supply.

Technological innovations are set to enhance the efficiency and sustainability of hydropower. Advances in turbine design, such as the development of fish-friendly and variable-speed turbines, are reducing environmental impacts while improving energy output. These innovations allow for more efficient energy capture from water flows, even in low-head or variable-flow conditions. Additionally, the integration of digital technologies and data analytics is enabling more precise monitoring and control of hydropower facilities, optimizing performance and reducing maintenance costs.

The modernization of existing hydropower infrastructure presents a significant opportunity for increasing capacity and efficiency without the need for new dam construction. Many hydropower plants around the world are aging and in need of upgrades. By retrofitting these facilities with modern equipment and control systems, it is possible to enhance their performance and extend their operational lifespan. This approach not only maximizes the use of existing resources but also minimizes the environmental and social impacts associated with new construction.

Pumped storage hydropower is another area of growth, offering a large-scale energy storage solution that complements the integration of renewable energy sources. By storing excess energy generated during periods of low demand and releasing it during peak demand, pumped storage systems provide a valuable means of balancing supply and demand. This capability is particularly important in grids with a high penetration of intermittent renewables, where energy storage is essential for maintaining grid stability.

Environmental sustainability is a critical consideration for the future of hydropower. While it is a clean energy source that does not produce direct emissions, the construction and operation of hydropower projects can have significant ecological effects. To address these challenges, comprehensive environmental impact assessments are conducted to identify and mitigate potential risks. The development of environmentally friendly technologies and practices, such as habitat restoration and adaptive management strategies, is helping to reduce the ecological footprint of hydropower projects.

Social considerations are equally important in the development of sustainable hydropower. The construction of dams and reservoirs can lead to the displacement of communities and changes in land use. Engaging with local stakeholders and ensuring that affected communities are involved in the decision-making process is crucial for the equitable development of hydropower projects. By prioritizing social responsibility and community engagement, it is possible to ensure that the benefits of hydropower are shared equitably and that negative impacts are minimized.

Policy and regulatory frameworks will play a vital role in shaping the future of hydropower. Governments can support the development of sustainable hydropower through financial incentives, such as grants, subsidies, and tax credits. Clear regulatory guidelines and streamlined permitting processes can facilitate the development of projects, reducing administrative barriers and encouraging innovation. International cooperation and knowledge sharing can further enhance the capacity of countries to implement sustainable hydropower projects.

The integration of hydropower into a sustainable energy mix will require a holistic approach that considers the interplay between different energy sources. By leveraging the complementary strengths of hydropower, wind, solar, and other renewables, it is possible to create a balanced and resilient energy system. Hybrid energy systems that combine multiple renewable sources can provide a more consistent and reliable energy supply, reducing reliance on fossil fuels and contributing to the reduction of greenhouse gas emissions.

The future of hydropower is promising, with the potential to play a key role in the transition to a sustainable energy future. Continued innovation and investment in research and development will be essential for overcoming existing challenges and unlocking new opportunities. By embracing these advancements, we can harness the full potential of hydropower to create a cleaner, more sustainable energy future for generations to come. As the world continues to transition towards renewable energy, the integration of hydropower into the global energy landscape will remain a vital component of the transition to a low-carbon economy.

Chapter 4: Geothermal Energy: Tapping into Earth's Core

Understanding Geothermal Energy Systems

Geothermal energy systems tap into the Earth's internal heat to provide a sustainable and reliable source of energy. This form of energy is derived from the natural heat stored beneath the Earth's surface, which can be harnessed for electricity generation, direct heating applications, and even cooling. Understanding how geothermal energy systems work and their potential benefits is crucial for anyone interested in renewable energy solutions.

The Earth's core generates heat through the natural decay of radioactive isotopes and the residual heat from the planet's formation. This heat gradually moves towards the surface, creating geothermal reservoirs of hot water and steam. These reservoirs can be found in regions with high tectonic activity, such as along the Pacific Ring of Fire, but they also exist in other areas with suitable geological conditions.

Geothermal energy systems can be broadly categorized into three main types: dry steam, flash steam, and binary cycle. Each type utilizes different methods to convert geothermal heat into usable energy, depending on the temperature and pressure of the geothermal resource.

Dry steam systems are the simplest and oldest form of geothermal energy technology. They directly use steam from geothermal reservoirs to drive turbines connected to electricity generators. The steam is extracted from underground wells and piped to the power plant, where it spins the turbines to produce electricity. After passing through the turbines, the steam is condensed back into water and reinjected into the reservoir to maintain pressure and sustainability. Dry steam systems are typically found in areas with high-temperature geothermal resources, such as the Geysers in California.

Flash steam systems are the most common type of geothermal power plant. They utilize geothermal reservoirs with water temperatures above 182°C (360°F). In these systems, high-pressure hot water is extracted from the ground and transported to the surface. As the pressure decreases, some of the water "flashes" into steam, which is then used to drive turbines and generate electricity. The remaining water and condensed steam are reinjected into the reservoir to sustain the resource. Flash steam systems are highly efficient and can be found in geothermal fields around the world, including Iceland and the Philippines.

Binary cycle systems are designed to utilize lower-temperature geothermal resources, typically between 107°C (225°F) and 182°C (360°F). In these systems, geothermal water is passed through a heat exchanger, where it transfers its heat to a secondary fluid with a lower boiling point, such as isobutane or pentane. The secondary fluid vaporizes and drives a turbine to generate electricity. The geothermal water is then reinjected into the reservoir, while the secondary fluid is condensed and reused in a closed-loop system. Binary cycle systems are versatile and can be deployed in a wide range of locations, making them an attractive option for expanding geothermal energy use.

Beyond electricity generation, geothermal energy systems offer a variety of direct-use applications. These include district heating, greenhouse heating, aquaculture pond heating, and industrial processes. Geothermal heat pumps, also known as ground-source heat pumps, are another popular application. They utilize the stable temperatures of the Earth's subsurface to provide heating and cooling for residential and commercial buildings. By circulating a fluid through underground pipes, geothermal heat pumps can transfer heat from the ground into buildings during the winter and remove heat from buildings during the summer, offering an energy-efficient alternative to traditional HVAC systems.

The environmental benefits of geothermal energy systems are significant. They produce minimal greenhouse gas emissions compared to fossil fuel-based power plants, contributing to the reduction of carbon footprints and the mitigation of climate change. Geothermal power plants also have a small land footprint, as they require less space than solar or wind farms. Additionally, the continuous and reliable nature of geothermal energy makes it an ideal complement to intermittent renewable sources, such as wind and solar, enhancing grid stability and energy security.

However, the development of geothermal energy systems is not without challenges. The exploration and drilling of geothermal wells can be costly and risky, as the presence and quality of geothermal resources are not always guaranteed. Advanced exploration techniques, such as seismic surveys and remote sensing, are being developed to improve the accuracy of resource assessments and reduce financial risks. Additionally, the potential for induced seismicity, or human-caused earthquakes, is a concern in some geothermal projects. Careful site selection, monitoring, and management practices are essential to minimize these risks and ensure the safe and sustainable development of geothermal resources.

Policy and regulatory support play a crucial role in the expansion of geothermal energy systems. Governments can encourage investment in geothermal projects through financial incentives, such as grants, tax credits, and low-interest loans. Streamlined permitting processes and clear regulatory frameworks can facilitate project development and reduce administrative barriers. International collaboration and knowledge sharing can further accelerate the adoption of best practices and technological advancements, driving the global expansion of geothermal energy.

The future of geothermal energy systems is promising, with the potential to play a key role in the transition to a sustainable energy future. Continued innovation and investment in research and development will be essential for overcoming existing challenges and unlocking new opportunities. By embracing these advancements, we can harness the full potential of geothermal energy to create a cleaner, more sustainable energy future for generations to come. As the world continues to transition towards renewable energy, the integration of geothermal energy into the global energy landscape will remain a vital component of the transition to a low-carbon economy.

Advantages and Challenges of Geothermal Adoption

Geothermal energy, a formidable player in the realm of renewable resources, offers a unique set of advantages and challenges that influence its adoption across the globe. As the world increasingly turns to sustainable energy solutions, understanding these facets of geothermal energy is essential for stakeholders, from policymakers to local communities.

One of the most compelling advantages of geothermal energy is its sustainability. Unlike fossil fuels, geothermal energy is derived from the Earth's internal heat, which is virtually inexhaustible on a human timescale. This makes it a reliable and long-term energy source that can significantly reduce carbon emissions and combat climate change. Geothermal power plants emit minimal greenhouse gases compared to coal or natural gas plants, making them an environmentally friendly option for electricity generation.

The reliability of geothermal energy is another significant advantage. Unlike solar and wind energy, which are dependent on weather conditions, geothermal energy provides a constant and stable power output. This baseload capability ensures a continuous supply of electricity, enhancing grid stability and reducing the need for backup power sources. This reliability is particularly valuable in regions with high energy demands or limited access to other renewable resources.

Geothermal energy systems also offer versatility in their applications. Beyond electricity generation, geothermal energy can be used for direct heating purposes, such as district heating, greenhouse heating, and industrial processes. Geothermal heat pumps provide efficient heating and cooling solutions for residential and commercial buildings, utilizing the stable temperatures of the Earth's subsurface. This versatility allows geothermal energy to meet a wide range of energy needs, from large-scale power generation to localized heating and cooling.

Economic benefits are another key advantage of geothermal energy adoption. The development and operation of geothermal projects can create jobs and stimulate local economies. From exploration and drilling to plant construction and maintenance, geothermal projects require a skilled workforce, providing employment opportunities in various sectors. Additionally, the use of domestic geothermal resources can reduce dependence on imported fuels, enhancing energy security and contributing to economic stability.

Despite these advantages, the adoption of geothermal energy faces several challenges that must be addressed to realize its full potential. One of the primary challenges is the high upfront cost associated with geothermal exploration and development. Identifying viable geothermal resources requires extensive geological surveys and drilling, which can be expensive and risky. The uncertainty of resource availability and quality adds to the financial risk, making it difficult for developers to secure funding and investment.

Technological challenges also play a role in geothermal energy adoption. While advances in drilling and exploration technologies have improved the efficiency and success rate of geothermal projects, there is still a need for further innovation to reduce costs and enhance resource utilization. Developing technologies that can tap into deeper and hotter geothermal resources could significantly expand the potential of geothermal energy, but these advancements require substantial research and development efforts.

Environmental and social considerations are important factors in the adoption of geothermal energy. While geothermal power plants have a relatively small environmental footprint, the development of geothermal resources can have localized impacts, such as land use changes and the potential for induced seismicity. Engaging with local communities and conducting thorough environmental impact assessments are essential to address these concerns and ensure the sustainable development of geothermal projects.

Policy and regulatory frameworks play a crucial role in facilitating geothermal energy adoption. Governments can support the development of geothermal projects through financial incentives, such as grants, tax credits, and low-interest loans. Streamlined permitting processes and clear regulatory guidelines can reduce administrative barriers and encourage investment. International collaboration and knowledge sharing can further enhance the capacity of countries to implement geothermal projects and adopt best practices.

Public perception and awareness are also critical to the successful adoption of geothermal energy. Educating communities about the benefits and potential impacts of geothermal projects can foster public support and acceptance. Transparent communication and stakeholder engagement are essential to building trust and addressing any concerns or misconceptions about geothermal energy.

The future of geothermal energy adoption is promising, with the potential to play a key role in the transition to a sustainable energy future. Continued innovation and investment in research and development will be essential for overcoming existing challenges and unlocking new opportunities. By embracing these advancements, we can harness the full potential of geothermal energy to create a cleaner, more sustainable energy future for generations to come. As the world continues to transition towards renewable energy, the integration of geothermal energy into the global energy landscape will remain a vital component of the transition to a low-carbon economy.

Global Leaders in Geothermal Development

Geothermal energy, a cornerstone of the renewable energy landscape, has seen significant advancements and adoption across the globe. Several countries have emerged as leaders in geothermal development, each contributing unique innovations and strategies to harness the Earth's heat. These global leaders not only demonstrate the potential of geothermal energy but also provide valuable lessons for other nations seeking to expand their renewable energy portfolios.

Iceland stands as a beacon of geothermal success, with nearly 90% of its homes heated by geothermal energy and a significant portion of its electricity generated from geothermal power plants. The country's unique geological conditions, characterized by abundant volcanic activity, provide an ideal environment for geothermal exploitation. Iceland's commitment to renewable energy is deeply rooted in its national policy, which prioritizes sustainability and energy independence. The country's expertise in geothermal technology and resource management has made it a global leader, offering consultancy and training services to other nations interested in developing their geothermal resources.

The United States, with its vast and diverse geothermal resources, is another prominent player in the geothermal arena. The western states, particularly California and Nevada, host some of the largest geothermal power plants in the world. The Geysers in California, for instance, is the largest complex of geothermal power plants globally. The U.S. government's support for geothermal research and development, coupled with private sector investment, has driven technological advancements and increased capacity. The country's focus on innovation has led to the development of enhanced geothermal systems (EGS), which have the potential to unlock geothermal resources in areas previously deemed unsuitable.

Kenya has emerged as a leader in geothermal development within Africa, leveraging its geothermal resources to address energy shortages and promote economic growth. The East African Rift Valley, which runs through Kenya, provides significant geothermal potential. The Olkaria Geothermal Plant, one of the largest in Africa, exemplifies Kenya's commitment to renewable energy. The Kenyan government's proactive approach, including favorable policies and investment incentives, has attracted international partnerships and funding. As a result, geothermal energy now accounts for a substantial portion of Kenya's electricity generation, reducing reliance on fossil fuels and enhancing energy security.

The Philippines is another notable leader in geothermal energy, ranking among the top producers worldwide. The country's location along the Pacific Ring of Fire provides abundant geothermal resources, which have been harnessed for electricity generation since the 1970s. The Philippine government's support for geothermal development, through policies and incentives, has fostered a thriving industry. The country's experience in geothermal exploration and development has positioned it as a regional leader, with expertise that is sought after by neighboring countries looking to develop their geothermal resources.

New Zealand, with its rich geothermal resources, has long been a pioneer in geothermal energy. The country was one of the first to develop geothermal power plants, and today, geothermal energy accounts for a significant portion of its electricity generation. New Zealand's approach to geothermal development emphasizes sustainability and environmental stewardship, ensuring that geothermal resources are managed responsibly. The country's expertise in geothermal technology and resource management has made it a global leader, contributing to international research and development efforts.

Japan, despite its limited land area, has made significant strides in geothermal development. The country's location along the Pacific Ring of Fire provides substantial geothermal potential, which is being increasingly tapped to diversify its energy mix. Japan's focus on innovation and technology has led to the development of advanced geothermal systems, including binary cycle power plants that can utilize lower-temperature resources. The Japanese government's support for renewable energy, particularly in the wake of the Fukushima nuclear disaster, has accelerated geothermal development and positioned the country as a leader in the field.

Italy, with its long history of geothermal energy use, remains a key player in the global geothermal landscape. The country is home to the world's first geothermal power plant, built in Larderello in 1911. Today, Italy continues to be a leader in geothermal technology and innovation, with a focus on sustainable development and environmental protection. The Italian government's support for geothermal research and development, along with favorable policies and incentives, has fostered a robust industry that contributes significantly to the country's energy mix.

These global leaders in geothermal development offer valuable insights and lessons for other countries seeking to harness their geothermal resources. Key factors contributing to their success include favorable geological conditions, supportive government policies, investment in research and development, and a commitment to sustainability. By learning from these leaders, other nations can develop strategies to overcome challenges and unlock the potential of geothermal energy.

Collaboration and knowledge sharing among countries are essential for advancing geothermal development worldwide. International partnerships, such as those facilitated by organizations like the International Geothermal Association, provide opportunities for countries to share expertise, technology, and best practices. These collaborations can accelerate the adoption of geothermal energy and contribute to the global transition to a sustainable energy future.

The future of geothermal energy is promising, with the potential to play a key role in the transition to a sustainable energy future. Continued innovation and investment in research and development will be essential for overcoming existing challenges and unlocking new opportunities. By embracing these advancements, we can harness the full potential of geothermal energy to create a cleaner, more sustainable energy future for generations to come. As the world continues to transition towards renewable energy, the integration of geothermal energy into the global energy landscape will remain a vital component of the transition to a low-carbon economy.

Technological Innovations in Geothermal Exploration

Geothermal exploration has long been a cornerstone of renewable energy development, offering a sustainable and reliable source of power. As the demand for clean energy grows, technological innovations in geothermal exploration are becoming increasingly important. These advancements are not only enhancing the efficiency and effectiveness of geothermal projects but also expanding the potential for geothermal energy in regions previously considered unsuitable.

One of the most significant technological innovations in geothermal exploration is the use of advanced geophysical techniques. These methods, including seismic surveys, magnetotellurics, and gravity measurements, allow for a more accurate assessment of geothermal resources. Seismic surveys, for instance, use sound waves to create detailed images of subsurface structures, helping to identify potential geothermal reservoirs. Magnetotellurics measures the Earth's natural electromagnetic fields to map subsurface conductivity, providing valuable information about the presence of hot fluids. Gravity measurements detect variations in the Earth's gravitational field, which can indicate the presence of geothermal resources. By combining these techniques, geologists can create comprehensive models of geothermal reservoirs, reducing the risk and uncertainty associated with exploration.

Another key innovation is the development of remote sensing technologies. Satellite imagery and aerial surveys provide valuable data on surface features and thermal anomalies, which can indicate the presence of geothermal activity. These technologies allow for large-scale assessments of geothermal potential, enabling the identification of promising sites for further exploration. Remote sensing is particularly useful in remote or inaccessible areas, where traditional exploration methods may be challenging or costly.

Drilling technology has also seen significant advancements, with the development of more efficient and cost-effective drilling techniques. Directional drilling, for example, allows for the precise targeting of geothermal reservoirs, reducing the need for multiple exploratory wells. This technique involves drilling at an angle, rather than vertically, to reach specific subsurface targets. Enhanced drilling fluids and materials have improved the durability and performance of drilling equipment, reducing downtime and maintenance costs. Additionally, the use of real-time data monitoring during drilling operations allows for more informed decision-making and adjustments, optimizing the drilling process and increasing the likelihood of success.

Enhanced geothermal systems (EGS) represent a groundbreaking innovation in geothermal exploration. Traditional geothermal systems rely on naturally occurring reservoirs of hot water and steam, which are limited to specific geological conditions. EGS, on the other hand, involves creating artificial reservoirs by injecting water into hot, dry rock formations. This process, known as hydraulic stimulation, enhances the permeability of the rock, allowing for the extraction of heat. EGS has the potential to significantly expand the availability of geothermal resources, making it possible to harness geothermal energy in regions without natural reservoirs. While still in the experimental stage, EGS projects have shown promising results, and continued research and development could unlock vast new geothermal potential.

The integration of digital technologies and data analytics is transforming geothermal exploration. Advanced software and modeling tools enable the analysis of complex geological data, providing insights into subsurface conditions and resource potential. Machine learning algorithms can identify patterns and correlations in large datasets, improving the accuracy of resource assessments and reducing exploration risks. These technologies also facilitate real-time monitoring and management of geothermal projects, optimizing performance and efficiency.

Collaboration and knowledge sharing are essential components of technological innovation in geothermal exploration. International partnerships and research initiatives bring together experts from diverse fields, fostering the exchange of ideas and best practices. Organizations such as the International Geothermal Association and the Geothermal Resources Council play a crucial role in facilitating collaboration and promoting technological advancements. By working together, countries can leverage their collective expertise to overcome challenges and accelerate the development of geothermal energy.

The role of government policy and support in driving technological innovation cannot be overstated. Governments can incentivize research and development through grants, subsidies, and tax credits, encouraging investment in new technologies. Clear regulatory frameworks and streamlined permitting processes can facilitate the deployment of innovative geothermal projects, reducing administrative barriers and fostering a favorable environment for exploration. Public-private partnerships can also play a vital role in advancing geothermal technology, combining the resources and expertise of both sectors to achieve common goals.

As technological innovations continue to reshape geothermal exploration, the potential for geothermal energy is expanding. These advancements are not only making geothermal projects more efficient and cost-effective but also opening up new opportunities for development in regions previously considered unsuitable. By embracing these innovations, the geothermal industry can play a pivotal role in the transition to a sustainable energy future.

The future of geothermal exploration is bright, with the potential to unlock vast new resources and contribute significantly to global energy needs. Continued investment in research and development, coupled with international collaboration and supportive government policies, will be essential for realizing this potential. By harnessing the power of technological innovation, we can create a cleaner, more sustainable energy future for generations to come.

Integrating Geothermal Energy into National Grids

Integrating geothermal energy into national grids presents a unique opportunity to enhance energy security, reduce carbon emissions, and provide a stable power supply. As countries strive to diversify their energy portfolios and transition to renewable sources, geothermal energy offers a reliable and sustainable solution. However, the integration process involves a series of technical, economic, and regulatory considerations that must be addressed to maximize the benefits of geothermal energy.

One of the primary advantages of geothermal energy is its ability to provide baseload power. Unlike solar and wind energy, which are intermittent and dependent on weather conditions, geothermal energy offers a constant and stable output. This reliability makes it an ideal candidate for integration into national grids, where a consistent power supply is essential for maintaining grid stability and meeting demand. By incorporating geothermal energy into the grid, countries can reduce their reliance on fossil fuels and enhance energy security.

The integration of geothermal energy into national grids requires careful planning and coordination. Grid operators must assess the capacity and potential of geothermal resources, taking into account factors such as location, resource temperature, and reservoir size. This assessment helps determine the feasibility of connecting geothermal power plants to the grid and ensures that the infrastructure can accommodate the additional capacity. In some cases, upgrades to transmission lines and substations may be necessary to support the integration of geothermal energy.

Economic considerations play a crucial role in the integration process. The cost of developing geothermal projects, including exploration, drilling, and plant construction, can be significant. However, the long-term benefits of geothermal energy, such as reduced fuel costs and lower emissions, often outweigh the initial investment. Governments can support the integration of geothermal energy by providing financial incentives, such as grants, tax credits, and low-interest loans, to offset the upfront costs and encourage investment in geothermal projects.

Regulatory frameworks are essential for facilitating the integration of geothermal energy into national grids. Clear and consistent policies can streamline the permitting process, reduce administrative barriers, and provide a stable environment for investment. Governments can establish feed-in tariffs or power purchase agreements to guarantee a market for geothermal energy and ensure that developers receive a fair return on their investment. Additionally, regulations should address environmental and social considerations, ensuring that geothermal projects are developed sustainably and with minimal impact on local communities.

Public perception and acceptance are critical to the successful integration of geothermal energy into national grids. Educating communities about the benefits of geothermal energy, such as reduced emissions and increased energy security, can foster public support and acceptance. Transparent communication and stakeholder engagement are essential for building trust and addressing any concerns or misconceptions about geothermal energy. By involving local communities in the planning and development process, governments and developers can ensure that geothermal projects are aligned with local needs and priorities.

Technological advancements are playing a significant role in facilitating the integration of geothermal energy into national grids. Innovations in drilling and exploration techniques are reducing the cost and risk associated with geothermal projects, making them more attractive to investors. Enhanced geothermal systems (EGS) are expanding the potential for geothermal energy by enabling the development of resources in areas without naturally occurring reservoirs. These advancements are increasing the availability and accessibility of geothermal energy, making it a viable option for integration into national grids.

International collaboration and knowledge sharing are essential for advancing the integration of geothermal energy into national grids. Countries can learn from the experiences of global leaders in geothermal development, such as Iceland, the United States, and Kenya, to identify best practices and strategies for successful integration. Organizations like the International Renewable Energy Agency (IRENA) and the International Geothermal Association provide platforms for collaboration and exchange, facilitating the dissemination of knowledge and expertise.

The integration of geothermal energy into national grids offers significant environmental benefits. By displacing fossil fuel-based power generation, geothermal energy can contribute to a reduction in greenhouse gas emissions and air pollution. This transition supports national and international climate goals, helping countries meet their commitments under agreements like the Paris Accord. Additionally, geothermal energy has a relatively small land footprint compared to other renewable sources, such as solar and wind, making it an attractive option for densely populated areas or regions with limited land availability.

The future of geothermal energy integration is promising, with the potential to play a key role in the transition to a sustainable energy future. Continued investment in research and development, coupled with supportive government policies and international collaboration, will be essential for realizing this potential. By harnessing the power of geothermal energy, countries can create a cleaner, more sustainable energy future for generations to come. As the world continues to transition towards renewable energy, the integration of geothermal energy into national grids will remain a vital component of the transition to a low-carbon economy.

Chapter 5: Bioenergy: Transforming Organic Matter into Power

Overview of Bioenergy Sources and Technologies

Bioenergy, a versatile and renewable energy source, plays a crucial role in the global transition towards sustainable energy systems. Derived from organic materials, bioenergy encompasses a wide range of sources and technologies that convert biomass into usable energy forms. Understanding the diversity of bioenergy sources and the technologies used to harness them is essential for effectively integrating bioenergy into national and regional energy strategies.

Biomass, the primary source of bioenergy, includes a variety of organic materials such as agricultural residues, forestry by-products, animal manure, and dedicated energy crops. Each type of biomass offers unique characteristics and energy potential, influencing the choice of conversion technology and end-use application. Agricultural residues, such as straw and corn stover, are abundant and readily available, making them attractive feedstocks for bioenergy production. Forestry by-products, including wood chips and sawdust, provide a sustainable source of biomass, particularly in regions with well-managed forests. Animal manure, rich in organic matter, can be converted into biogas through anaerobic digestion, offering a valuable energy source for rural communities. Dedicated energy crops, such as switchgrass and miscanthus, are specifically cultivated for bioenergy production, providing high yields and consistent quality.

The conversion of biomass into bioenergy involves a range of technologies, each suited to specific feedstocks and energy applications. Combustion, one of the oldest and most straightforward technologies, involves burning biomass to produce heat and power. This process is commonly used in industrial and district heating applications, where large-scale biomass boilers provide a reliable and efficient energy source. Co-firing, a variation of combustion, involves burning biomass alongside coal in existing power plants, reducing carbon emissions and extending the life of coal-fired facilities.

Gasification, a more advanced technology, converts biomass into a combustible gas mixture known as syngas. This process involves heating biomass in a low-oxygen environment, resulting in the production of syngas, which can be used for electricity generation, heating, or as a feedstock for chemical synthesis. Gasification offers higher efficiency and lower emissions compared to traditional combustion, making it an attractive option for bioenergy production.

Anaerobic digestion, a biological process, breaks down organic matter in the absence of oxygen to produce biogas, a mixture of methane and carbon dioxide. This technology is particularly suited to wet biomass feedstocks, such as animal manure, food waste, and sewage sludge. Biogas can be used for electricity and heat generation or upgraded to biomethane, a renewable natural gas substitute suitable for injection into natural gas grids or use as vehicle fuel.

Pyrolysis, another thermochemical process, involves heating biomass in the absence of oxygen to produce bio-oil, syngas, and biochar. Bio-oil, a liquid fuel, can be used for heating or further refined into transportation fuels. Biochar, a carbon-rich solid, can be used as a soil amendment, improving soil fertility and sequestering carbon. Pyrolysis offers a flexible and efficient method for converting biomass into multiple valuable products.

Fermentation, a biochemical process, converts sugars and starches in biomass into ethanol, a renewable liquid fuel. This technology is widely used in the production of biofuels from crops such as corn, sugarcane, and wheat. Ethanol can be blended with gasoline to reduce emissions and enhance fuel performance. Advances in fermentation technology are enabling the production of cellulosic ethanol from non-food biomass, such as agricultural residues and dedicated energy crops, reducing competition with food production and enhancing sustainability.

Biodiesel production, another key bioenergy technology, involves the transesterification of vegetable oils or animal fats to produce biodiesel, a renewable diesel substitute. Biodiesel can be used in existing diesel engines with little or no modification, offering a sustainable alternative to fossil diesel. Feedstocks for biodiesel production include soybean oil, rapeseed oil, and used cooking oil, providing a diverse and flexible resource base.

The integration of bioenergy into energy systems offers numerous environmental and economic benefits. By displacing fossil fuels, bioenergy can significantly reduce greenhouse gas emissions and air pollution, contributing to climate change mitigation and improved air quality. The use of locally sourced biomass can enhance energy security and reduce dependence on imported fuels, supporting regional economic development and job creation. Bioenergy also offers opportunities for waste management and resource recovery, turning organic waste streams into valuable energy products.

Despite its potential, the deployment of bioenergy faces several challenges that must be addressed to realize its full benefits. The availability and sustainability of biomass feedstocks are critical considerations, requiring careful management and planning to ensure a reliable and consistent supply. Land use and competition with food production are important factors, necessitating the development of sustainable biomass sourcing strategies that balance energy production with food security and environmental protection.

Technological advancements and innovation are essential for overcoming these challenges and enhancing the efficiency and sustainability of bioenergy systems. Continued research and development in areas such as feedstock improvement, conversion technology optimization, and integrated bioenergy systems will be crucial for advancing the bioenergy sector. Collaboration and knowledge sharing among stakeholders, including governments, industry, and research institutions, can facilitate the exchange of best practices and drive the adoption of innovative solutions.

Policy and regulatory frameworks play a vital role in supporting the development and deployment of bioenergy. Governments can incentivize bioenergy production through financial support mechanisms, such as grants, subsidies, and tax credits, as well as by establishing renewable energy targets and mandates. Clear and consistent regulations can provide a stable environment for investment and ensure that bioenergy projects are developed sustainably and responsibly.

Public perception and acceptance are also important factors in the successful deployment of bioenergy. Educating communities about the benefits and potential impacts of bioenergy can foster public support and acceptance. Transparent communication and stakeholder engagement are essential for building trust and addressing any concerns or misconceptions about bioenergy.

The future of bioenergy is promising, with the potential to play a key role in the transition to a sustainable energy future. By harnessing the diverse range of bioenergy sources and technologies, countries can create a cleaner, more sustainable energy system that supports economic growth, environmental protection, and energy security. Continued investment in research and development, coupled with supportive policies and international collaboration, will be essential for realizing this potential and unlocking the full benefits of bioenergy.

The Role of Bioenergy in Circular Economies

Bioenergy, a pivotal component of renewable energy strategies, plays a significant role in advancing circular economies. By transforming organic waste into valuable energy resources, bioenergy not only contributes to sustainable energy production but also supports waste management and resource efficiency. Understanding the role of bioenergy in circular economies is essential for developing integrated systems that maximize resource use and minimize environmental impact.

Circular economies aim to create closed-loop systems where resources are reused, recycled, and regenerated, reducing waste and conserving natural resources. Bioenergy aligns with these principles by converting organic waste streams into energy, thereby closing the loop on resource use. This process not only reduces the volume of waste sent to landfills but also recovers valuable energy, contributing to a more sustainable and resilient energy system.

One of the primary ways bioenergy supports circular economies is through the utilization of agricultural and forestry residues. These by-products, often considered waste, can be converted into bioenergy, providing a renewable energy source while reducing the need for disposal. For example, straw, corn stover, and wood chips can be used as feedstocks for bioenergy production, turning waste into a valuable resource. This approach not only enhances resource efficiency but also provides additional income streams for farmers and foresters, supporting rural economies.

Animal manure, another abundant waste stream, can be transformed into biogas through anaerobic digestion. This process not only generates renewable energy but also produces nutrient-rich digestate, which can be used as a natural fertilizer, closing the nutrient loop in agricultural systems. By integrating bioenergy production with agricultural practices, farmers can reduce their reliance on synthetic fertilizers, lower greenhouse gas emissions, and improve soil health, contributing to more sustainable and resilient farming systems.

Food waste, a significant contributor to global waste streams, offers another opportunity for bioenergy production. Through anaerobic digestion or composting, food waste can be converted into biogas or biofertilizer, reducing landfill waste and recovering valuable resources. Municipalities and businesses can implement food waste collection and processing systems to harness this potential, supporting circular economy goals and reducing the environmental impact of waste disposal.

The role of bioenergy in circular economies extends beyond waste management to include the production of bio-based products and materials. By using biomass as a feedstock, industries can produce biofuels, bioplastics, and other bio-based materials, reducing reliance on fossil fuels and non-renewable resources. This approach not only supports the transition to a low-carbon economy but also fosters innovation and economic growth in the bioeconomy sector.

Technological advancements are driving the integration of bioenergy into circular economies, enhancing the efficiency and sustainability of bioenergy systems. Innovations in conversion technologies, such as advanced fermentation and gasification processes, are enabling the production of high-value bio-based products from diverse biomass feedstocks. These technologies are expanding the potential for bioenergy production and supporting the development of integrated biorefineries, which produce multiple products from a single feedstock, maximizing resource use and minimizing waste.

Collaboration and knowledge sharing are essential for advancing the role of bioenergy in circular economies. By working together, stakeholders from government, industry, and academia can develop integrated solutions that leverage the synergies between bioenergy and circular economy principles. International partnerships and research initiatives can facilitate the exchange of best practices and drive the adoption of innovative technologies and approaches.

Policy and regulatory frameworks play a crucial role in supporting the integration of bioenergy into circular economies. Governments can incentivize bioenergy production and the use of bio-based products through financial support mechanisms, such as grants, subsidies, and tax credits. Clear and consistent regulations can provide a stable environment for investment and ensure that bioenergy projects are developed sustainably and responsibly. By aligning policies with circular economy goals, governments can create a supportive environment for the growth of the bioenergy sector and the transition to a more sustainable economy.

Public perception and acceptance are important factors in the successful integration of bioenergy into circular economies. Educating communities about the benefits and potential impacts of bioenergy can foster public support and acceptance. Transparent communication and stakeholder engagement are essential for building trust and addressing any concerns or misconceptions about bioenergy. By involving local communities in the planning and development process, governments and developers can ensure that bioenergy projects are aligned with local needs and priorities.

The future of bioenergy in circular economies is promising, with the potential to play a key role in the transition to a sustainable energy future. By harnessing the diverse range of bioenergy sources and technologies, countries can create a cleaner, more sustainable energy system that supports economic growth, environmental protection, and resource efficiency. Continued investment in research and development, coupled with supportive policies and international collaboration, will be essential for realizing this potential and unlocking the full benefits of bioenergy in circular economies.

Environmental and Economic Benefits of Bioenergy

Bioenergy, derived from organic materials, offers a compelling solution to the dual challenges of environmental sustainability and economic development. As the world grapples with the impacts of climate change and the need for sustainable energy sources, bioenergy emerges as a key player in the transition to a low-carbon economy. By understanding the environmental and economic benefits of bioenergy, we can better appreciate its potential to transform energy systems and support sustainable development.

One of the most significant environmental benefits of bioenergy is its potential to reduce greenhouse gas emissions. Unlike fossil fuels, which release carbon dioxide that has been stored underground for millions of years, bioenergy is part of the current carbon cycle. When biomass is converted into energy, the carbon dioxide released is roughly equivalent to the amount absorbed by the plants during their growth. This closed-loop system helps to mitigate the impact of energy production on climate change. By displacing fossil fuels with bioenergy, countries can significantly reduce their carbon footprint and contribute to global efforts to limit temperature rise.

Bioenergy also offers a solution to the problem of waste management. Organic waste streams, such as agricultural residues, forestry by-products, and food waste, can be converted into valuable energy resources, reducing the volume of waste sent to landfills. This not only minimizes the environmental impact of waste disposal but also recovers valuable resources that would otherwise be lost. For example, anaerobic digestion of food waste produces biogas, a renewable energy source, and digestate, a nutrient-rich fertilizer. By integrating bioenergy production with waste management systems, municipalities and businesses can enhance resource efficiency and reduce environmental pollution.

The use of bioenergy can lead to improvements in air quality, particularly in regions where traditional biomass burning is common. In many developing countries, households rely on open fires or simple stoves for cooking and heating, leading to indoor air pollution and associated health problems. Modern bioenergy technologies, such as biogas digesters and efficient biomass stoves, offer cleaner alternatives that reduce emissions of harmful pollutants. By promoting the adoption of these technologies, governments can improve public health and reduce the burden of respiratory diseases.

Bioenergy contributes to biodiversity conservation by providing an alternative to deforestation and land degradation. In regions where wood is the primary source of energy, the demand for fuelwood can lead to unsustainable harvesting practices and habitat loss. By developing sustainable bioenergy systems that utilize agricultural residues, dedicated energy crops, and other non-forest biomass, countries can reduce pressure on natural forests and promote biodiversity conservation. Additionally, the cultivation of energy crops on marginal or degraded lands can enhance soil health and prevent erosion, further supporting ecosystem resilience.

The economic benefits of bioenergy are equally compelling, offering opportunities for job creation, rural development, and energy security. The bioenergy sector encompasses a wide range of activities, from feedstock production and collection to conversion and distribution, creating jobs across the value chain. In rural areas, bioenergy projects can provide additional income streams for farmers and landowners, supporting economic diversification and resilience. By investing in bioenergy infrastructure and technology, countries can stimulate economic growth and create new markets for bio-based products and services.

Bioenergy enhances energy security by diversifying energy sources and reducing dependence on imported fuels. By utilizing locally available biomass resources, countries can reduce their vulnerability to fluctuations in global energy markets and enhance their energy independence. This is particularly important for regions with limited access to fossil fuel resources or those seeking to transition away from coal and oil. By integrating bioenergy into national energy strategies, governments can build more resilient and sustainable energy systems that are less susceptible to external shocks.

The development of bioenergy can drive innovation and technological advancement, fostering the growth of the bioeconomy. Research and development in bioenergy technologies, such as advanced biofuels, biorefineries, and bio-based materials, can lead to the creation of new products and processes that enhance resource efficiency and sustainability. By supporting innovation in the bioenergy sector, governments and industry can unlock new opportunities for economic growth and competitiveness.

Despite its potential, the deployment of bioenergy faces several challenges that must be addressed to realize its full benefits. The availability and sustainability of biomass feedstocks are critical considerations, requiring careful management and planning to ensure a reliable and consistent supply. Land use and competition with food production are important factors, necessitating the development of sustainable biomass sourcing strategies that balance energy production with food security and environmental protection.

Policy and regulatory frameworks play a vital role in supporting the development and deployment of bioenergy. Governments can incentivize bioenergy production through financial support mechanisms, such as grants, subsidies, and tax credits, as well as by establishing renewable energy targets and mandates. Clear and consistent regulations can provide a stable environment for investment and ensure that bioenergy projects are developed sustainably and responsibly.

Public perception and acceptance are also important factors in the successful deployment of bioenergy. Educating communities about the benefits and potential impacts of bioenergy can foster public support and acceptance. Transparent communication and stakeholder engagement are essential for building trust and addressing any concerns or misconceptions about bioenergy.

The future of bioenergy is promising, with the potential to play a key role in the transition to a sustainable energy future. By harnessing the diverse range of bioenergy sources and technologies, countries can create a cleaner, more sustainable energy system that supports economic growth, environmental protection, and energy security. Continued investment in research and development, coupled with supportive policies and international collaboration, will be essential for realizing this potential and unlocking the full benefits of bioenergy.

Challenges in Scaling Bioenergy Solutions

Scaling bioenergy solutions presents a complex array of challenges that must be navigated to fully harness the potential of this renewable energy source. While bioenergy offers significant environmental and economic benefits, its widespread adoption requires overcoming technical, logistical, and policy-related hurdles. Understanding these challenges is crucial for developing effective strategies to expand bioenergy deployment and integrate it into national and global energy systems.

One of the primary challenges in scaling bioenergy solutions is the availability and sustainability of biomass feedstocks. Biomass, the raw material for bioenergy production, encompasses a diverse range of organic materials, including agricultural residues, forestry by-products, and dedicated energy crops. Ensuring a reliable and consistent supply of biomass is essential for the viability of bioenergy projects. However, the availability of biomass can be influenced by factors such as seasonal variations, competition with food production, and land use constraints. Developing sustainable biomass sourcing strategies that balance energy production with food security and environmental protection is critical for scaling bioenergy solutions.

Logistical challenges also play a significant role in the scalability of bioenergy. The collection, transportation, and storage of biomass feedstocks can be complex and costly, particularly in regions with dispersed or remote biomass resources. Efficient supply chain management is essential to minimize costs and ensure the timely delivery of biomass to conversion facilities. Innovations in logistics, such as the development of decentralized processing facilities and the use of advanced transportation technologies, can help address these challenges and enhance the scalability of bioenergy solutions.

Technological advancements are crucial for overcoming the challenges associated with bioenergy conversion processes. While traditional technologies, such as combustion and anaerobic digestion, are well-established, there is a need for continued research and development to improve efficiency and reduce costs. Advanced conversion technologies, such as gasification and pyrolysis, offer the potential for higher efficiency and greater flexibility in feedstock use. However, these technologies often require significant capital investment and technical expertise, which can be barriers to widespread adoption. Supporting innovation and technology transfer is essential for scaling bioenergy solutions and enhancing their competitiveness with other renewable energy sources.

Economic considerations are another critical factor in scaling bioenergy solutions. The cost of developing and operating bioenergy projects can be significant, particularly in comparison to other renewable energy sources such as solar and wind. Financial incentives, such as grants, subsidies, and tax credits, can help offset the upfront costs and encourage investment in bioenergy projects. Additionally, establishing stable and predictable markets for bioenergy products, such as biofuels and biogas, is essential for attracting investment and ensuring the long-term viability of bioenergy solutions.

Policy and regulatory frameworks play a vital role in supporting the scaling of bioenergy solutions. Clear and consistent policies can provide a stable environment for investment and ensure that bioenergy projects are developed sustainably and responsibly. Governments can establish renewable energy targets and mandates to drive the adoption of bioenergy and create a level playing field with other renewable energy sources. Additionally, regulations should address environmental and social considerations, ensuring that bioenergy projects are developed in a manner that minimizes negative impacts and maximizes benefits for local communities.

Public perception and acceptance are important factors in the successful scaling of bioenergy solutions. Educating communities about the benefits and potential impacts of bioenergy can foster public support and acceptance. Transparent communication and stakeholder engagement are essential for building trust and addressing any concerns or misconceptions about bioenergy. By involving local communities in the planning and development process, governments and developers can ensure that bioenergy projects are aligned with local needs and priorities.

International collaboration and knowledge sharing are essential for advancing the scaling of bioenergy solutions. By working together, countries can learn from each other's experiences and identify best practices for overcoming common challenges. Organizations such as the International Renewable Energy Agency (IRENA) and the Global Bioenergy Partnership provide platforms for collaboration and exchange, facilitating the dissemination of knowledge and expertise. By leveraging international partnerships, countries can accelerate the development and deployment of bioenergy solutions and contribute to global efforts to transition to a sustainable energy future.

Environmental considerations are also critical in scaling bioenergy solutions. While bioenergy offers significant environmental benefits, such as reducing greenhouse gas emissions and improving waste management, it is essential to ensure that bioenergy projects are developed sustainably. This includes minimizing land use impacts, protecting biodiversity, and ensuring that biomass sourcing does not lead to deforestation or habitat loss. Developing and implementing sustainability criteria and certification schemes can help ensure that bioenergy projects meet environmental standards and contribute to sustainable development goals.

The integration of bioenergy into existing energy systems presents additional challenges. Bioenergy must be compatible with existing infrastructure and technologies, such as power grids and transportation networks, to ensure seamless integration and maximize its potential. This may require upgrades to infrastructure, such as the development of biogas injection facilities for natural gas grids or the adaptation of power plants for co-firing with biomass. Additionally, bioenergy must be integrated into energy planning and policy frameworks to ensure that it complements other renewable energy sources and contributes to a balanced and resilient energy system.

The future of bioenergy is promising, with the potential to play a key role in the transition to a sustainable energy future. By addressing the challenges associated with scaling bioenergy solutions, countries can unlock the full potential of this renewable energy source and create a cleaner, more sustainable energy system. Continued investment in research and development, coupled with supportive policies and international collaboration, will be essential for realizing this potential and ensuring that bioenergy contributes to global efforts to combat climate change and promote sustainable development.